THE RELIABILITY OF MECHANICAL SYSTEMS

The reliability of mechanical systems

Edited by
JOHN DAVIDSON, MBA, BSc(Hons), CEng, FIMechE,
FIPlantE

IMechE Guides for the Process Industries

Published by
Mechanical Engineering Publications Limited for
The Institution of Mechanical Engineers
LONDON

First Published 1988

British Library Cataloguing in Publication Data

The Reliability of mechanical systems.
1. Engineering equipment. Reliability
I. Davidson, J. (John), 1929–
II. Series
620'.00452
ISBN 0 85298 675 0

Typeset by Paston Press, Loddon, Norfolk
Printed by Acolortone Ltd., Ipswich, Suffolk

CONTENTS

PREFACE

Process plants are becoming larger and more sophisticated, with increasing use of automated computer controls, and with consequent increases, by orders of magnitude, in capital cost and the complexity of the design process. In these circumstances how can the owners and operators of such plants, at the time that they need to authorise the expenditure of capital, be assured that the plant proposed will start-up, and operate, safely and efficiently, with high levels of availability, to produce the forecast rate of return on the capital employed? Many companies have tasted the bitter fruit of long, painful start-ups, low plant availability, and disastrous returns on capital. Yet, despite all the efforts and work of a dedicated but relatively small band of specialist Reliability Engineers, the great majority of practising engineers remain relatively ignorant of the basic concepts and philosophy of Reliability Engineering, as well as the reliability assessment techniques used for the prediction of the reliability of new process plant designs.

Part of the answer to this lies in the fact that Reliability Engineering is seen by the layman engineer as a largely mathematically-based, esoteric, applied science, and that the techniques used deal with the mathematics of uncertainty (probability, predictions, and so on) and a host of statistical techniques normally the prerogative of mathematicians. For many engineers this very mathematical basis tends to create a mental barrier to both a basic understanding of the concepts and techniques, and the desire to get involved in their use and implementation. As is so often the case in the applied sciences, while the theoretical background can be difficult to comprehend, the practise and application of the techniques can be relatively easy to understand in the great majority of cases. Practise can then lead to a better and deeper understanding of the theoretical background. This is certainly the case with Reliability Engineering, and the approach should be to get 'one's feet wet' as soon as possible by using Reliability Engineering techniques (such as the Weibull graphs) and let the theory catch up with the practice later.

This book, then, is not a specialist theoretical treatise on reliability. It is a practical guide aimed at those thousands of practising engineers who may have a general knowledge and passing understanding of the concepts of reliability, but who lack, or have forgotten, the precise understanding of the language of Reliability Engineering to be able to confidently make effective practical use of the techniques involved. The aim is to provide a basic statement of philosophy, principles, and concepts, and then to introduce the reader to the techniques and methodology of reliability assessment and evaluation as applied to mechanical (as opposed to electrical/electronic) components, equipment, and mechanical plant systems and sub-systems. A comparison of the problems associated with mechanical, as opposed to electrical/electronic, reliability is made in order to bring out the complex nature and lack of reliable failure data documentation available in the area of mechanical engineering. Guidance is also given in the area of the design of data collection systems, as, without component and equipment failure data, relevant to the specific problem area, analysis is limited by the relatively poor availability of published failure-rate data for mechanical equipment.

The benefits to be gained by the reader are as follows:

(a) For mechanical engineers in the process industries: alerting them to the need for reliability assessment and evaluation at all stages in the plant systems design from the conceptual design stage, down to the detailed mechanical design stage, in order that realistic predictions of likely plant availability of alternative design proposals can be made, and the necessary corrective action can be taken if these fail to meet expectations. A further vital benefit is to alert designers to the need to design data collection systems early in the design process in order that adequate and meaningful data is available at the right time in order to check and analyse performance against the predictions of plant suppliers

(b) For mechanical engineers in the engineering manufacturing industry: alerting them to the economic, commercial, technical and safety implications of mechanical reliability with a view to their designing, manufacturing, and selling products which are more competitive, safe, and cost effective.

(c) For students of mechanical engineering: as most of the existing academic publications lean heavily toward a theoretical approach aimed largely at electronic equipment or components, with an assumed constant failure rate, students gain a relatively narrow appreciation of the problems existing in the real world of mechanical engineering. This book should help to enhance existing academic publications and provide some 'flesh on the bones' of real life Reliability Engineering. It could possibly suggest areas where final year students may wish to carry out research projects.

ACKNOWLEDGEMENTS

The basic material comprising the bulk of this book was provided by a Reliability Working Party set up by the Process Industries Division of the Institution of Mechanical Engineers. The composition of this Working Party is as follows.

T. R. Moss (Chairman)	RM Consultants, Abingdon	Author
E. A. Boxall	CEGB, Barnwood	Author
J. O. Catchpole	Bechtel GB, London	Author
M. J. Harris	Manchester University	Author
P. Kirrane	Rex, Thompson, and Partners, Farnham	Author
Dr R. F. de la Mare	Bradford University	Author
C. Musgrave	British Gas Corporation, Hinkley	Author
D. W. Newton	Birmingham University	Author
Professor A. D. S. Carter	Consultant, Oxford	Member
P. Martin	Liverpool University	Member

The editor wishes to thank all members of the Working Party for all the time and effort which went into the preparation and discussion of the base material, and in particular to thank the Chairman and the Authors for the friendly reception and willing cooperation during the preparation of the final text.

BIOGRAPHICAL NOTES

T. R. Moss

Bob Moss is Chairman of the Mechanical Reliability Committee and Managing Director of RM Consultants Ltd, a company specialising in safety, reliability, and risk analysis of process plant. He has been involved in reliability work for over 20 years and was a founder member of the UKAEA Systems Reliability Service and the European Reliability Data Bank Association. Particular interests are mechanical reliability research, plant availability assessment, and reliability information systems.

E. A. Boxall

Ted Boxall has been a member of the Reliability Engineering Group of the CEGB Generation Development and Construction Division since 1979. He was previously concerned with the control and instrumentation of power station systems for many years. His interests include reliability data collection and analysis, and the study of the reliability and availability of power plant using analytical and simulation techniques.

J. O. Catchpole

John Catchpole joined Bechtel Limited in 1986 as Chief Reliability Engineer, with corporate responsibility for reliability on company projects, also offering specialist services in design and plant availability. He has over 20 years experience in reliability engineering. He has had specific responsibilities for developing and coordinating the design reliability programmes for on and offshore projects, reliability design appraisals, and the design and development of computerised plant failure capture and analysis systems.

M. J. Harris

John Harris is a Senior Lecturer in Manchester University's Department of Engineering where he is a member of the Nuclear Engineering Group, which specialises in research related to the safety of nuclear reactors. In recent years his own interests have been in various aspects of the methodology of probabilistic safety assessment and, hence, of plant reliability assessment. He has also co-authored a textbook on the not unrelated topic of plant maintenance management.

P. P. Kirrane

Paddy Kirrane is a Senior Consultant with Rex, Thompson, and Partners, who have provided reliability consultancy services to Government Departments and commercial concerns since 1965. Prior to joining RTP in 1981, he spent two years at A&AEE Boscombe Down where he first became involved in reliability and safety work on aircraft flight control systems. In recent years at RTP he has specialised in reliability studies on mechanical equipment in the defence and transport industries.

Dr R. F. de la Mare

Roger de la Mare is Chairman of Technological Management at the University of Bradford Management Centre and Managing Director of Technological Management Consultants Ltd, a company specialising in technology transfer, training, and research.

As a chartered professional engineer with a formal education and training in business, he worked for Courtaulds, Esso, and the Rio Tinto Zinc Corporation before joining academe, where his major interests in mechanical reliability involve capital asset management, reliability economics, and maintenance.

C. Musgrave

Clive Musgrave is the secretary of the IMechE Mechanical Reliability Committee and is a Chartered Engineer employed in the hazard assessment and reliability department of British Gas Plc (Headquarters).

His work covers all aspects of reliability engineering as applied to the equipment in use throughout the gas industry. Particular interests include data collection and analysis for gas pumping machinery, reliability prediction and assessment for pneumatic equipment, and mathematical modelling of systems.

D. W. Newton

David Newton served an engineering apprenticeship with Smith Industries. He studied mechanical engineering at North Gloucestershire Technical College and obtained an external BSc(Eng) from London University. After a further period in industry he was awarded an MSc in Quality and Reliability Engineering from Birmingham University. He has been a lecturer in the Department of Engineering Production at Birmingham since 1973. He is currently Course Director of the Postgraduate Programme in Engineering Production and Management. Particular interests include statistical aspects of quality assurance and reliability, on which he has lectured extensively in the UK and overseas. He has been Chairman of the Business and Industrial Section of the Royal Statistical Society, and for 1987–88 is a Vice President of the Society.

Professor A. D. S. Carter

Denis Carter is currently employed as a private consultant in reliability engineering. From 1957 to 1982 he was Head of the Mechanical Engineering Department at the Royal Military College of Science where he specialised in reliability work on armoured fighting vehicles, guns, and other military equipment for which a good deal of field data was available. He is the author of many papers on various aspects of the gas turbine and mechanical reliability. He has just published a new and extended edition of his book *Mechanical Reliability*. He has also chaired numerous government technical committees in areas of his own technical expertise.

P. Martin

Peter Martin is presently Senior Lecturer in Mechanical Engineering and Deputy Director of the School of Engineering Science at the University of Liverpool. His early career was spent with Rolls-Royce Ltd on aircraft engine design, and he was subsequently involved in engineering design consultancy. Present research interests are centred on the application of reliability engineering considerations during product design and their application to advanced manufacturing systems.

'Rem tene; verba sequentur'

Grasp the subject; the words will follow

CAIUS JULIUS VICTOR

PART ONE

The philosophy, principles, and concepts of reliability engineering

CHAPTER 1. INTRODUCTION

Most people have a clear idea of what reliability means. A reliable person is someone who can be depended upon to do the right thing, at the right time, works consistently, keeps his appointments, and finishes his work tasks on time, backed up by appropriate timely reports. In the same way a reliable piece of equipment is understood to be basically sound, to be able to meet its design specification, and give trouble-free performance in a given environment. Thus, a washing machine designed for an industrial/commercial environment is expected to be more robust than one for domestic use. The modern housewife or layman contemplating the purchase of a major domestic electrical appliance, a car, or a personal computer, say, may well consult such popular publications as *Which?*, *Which Car?*, *Which Computer?*, and so on, to seek guidance on its 'reliability' as well as a whole host of other factors which are taken into account when evaluations are made for recommended best buys. The clear understanding of reliability on the part of the layman is based on the dictionary definitions of reliable (that is, that may be relied upon, of sound and consistent character or quality) and reliability trials (that is, long-distance trials, designed to test dependability, endurance, and so on).

In this book, however, we are concerned about a clear understanding of the technical, engineering, use of the term 'reliability'. We are concerned about the understanding of the technical concepts of reliability engineering. Engineers and managers must realize and clearly understand that in the real world nothing is perfect, that all plant, equipment, and components have a finite life, and that eventually even the very best of plant, equipment, or components will fail. Without a technical definition of reliability to which numerical values can be allocated it would not be possible for engineers or managers to make meaningful comparisons between the reliability of alternative plant and equipment proposals, it would not be possible for detailed analysis of component failures to point the way toward reliability improvements, and the science of reliability engineering would not exist. It is vitally important, therefore, that in the world of reliability engineering, all concerned clearly understand that reliability means a statement of the probability (a statistical concept) that a plant, piece of equipment, or component will not fail in a given time while working in a stated environment. This probability is given either a percentage value or a numerical value of 1 or less than 1. Thus, if the question were asked: 'What is the reliability of a seal on the methanol pump of the XYZ plant?' The possible answers could be categorised in three ways, as follows.

(a) 100 per cent (or 1) – the seal will certainly not fail in the given time period.

(b) Less than 100 per cent, but greater than 0 per cent – there is a defined chance (for example, 50 per cent) that the seal will fail in the time period.

(c) 0 per cent – the seal will certainly fail in the time period.

A full definition of mechanical reliability is discussed in section 1.2.

There is clearly a need to have a basic understanding of the mathematics of uncertainty, of how the probabilities of failure of individual components affect the probability of failure of a piece of equipment within which they are linked together, and of the language used in the world of reliability engineering. While the intention is to keep mathematical analysis to a minimum, and concentrate on methodology and techniques, it is impossible to avoid some mathematics, if only to explain the definition of terms and the interlinking of some of the basic concepts.

It is also important for the engineer or manager to realise that there are significant differences in the nature of the problems associated with the reliability of mechanical plant, equipment, and components, compared with electronic items. These differences will be spelled out, and it will be demonstrated how such differences should be accommodated in the search for improved design, operation, and maintenance.

Finally, it is important to realize that reliability analysis and prediction, of itself, will not solve the problems of unreliability and poor availability. It will only provide information to form a basis for either further engineering investigation or rational decisions on whether to modify or replace components, re-design the plant system, or increase the levels of certain items of spare (redundant) plant.

1.1 The implications of mechanical reliability

The practice of engineering is, perhaps surprisingly, like many other facets of life, subject to the whims of fashion. There have been 'fashions' of value engineering, hazard and operability studies, project task force teams, computerised maintenance management, and CAD, among numerous others. Each of these fashions can be hailed as the saviour of engineering and an almost universal panacea for all problems. It is important that a realistic, balanced view of the place of reliability is the order of things.

Reliability is only one of the many attributes to be taken into account in the overall design of components, equipment, or process plants. Other features such as weight, size, payload, throughput, yield, fuel economy, capital cost, and life cycle costs, as well as physical parameters which affect appearance and timeliness, may be more important to the final design than the achievement of the highest possible reliability. A strategy may be adopted to get a product on the market first to beat the competition and then to tackle its reliability as a second stage priority. The designers task is to achieve an overall design which is optimal in fulfilling the 'fitness for purpose' of the particular product. This usually requires a compromise between attributes because they can be mutually incompatible. For example mechanical reliability often increases with robustness, achieved by making the product more rigid by the use of thicker materials or the introduction of webs, and so on, but the extra weight or cost which results might affect the product speed, performance, or competitiveness. The reader must bear in mind, therefore, that while the emphasis here is on a product's reliability, many other needs, constraints, and priorities will effect the ultimate reliability that is designed and built into it.

(a) Economics

Since the early 1960s there has been a revolution in consumer buying habits, with consumers becoming more sophisticated in their purchasing decisions. Evidence suggests that, in many cases, the 'cheapest' product is no longer good enough, and the consumer's motto is increasingly tending toward 'value for money'. As mentioned earlier, the increasing proliferation of consumer magazines, such as *Which?*, and the annual AA Guide to Motoring Costs, provides ample evidence that consumers are looking beyond the first cost of a product, and considering the other costs which appertain throughout its life, that is, its Life Cycle Costs (LCC). In other words, besides first costs, they also consider the product's likely operating, maintenance, and replacement costs, coupled with the inconvenience caused by its partial or complete breakdown due to unreliability. A prime example of this is the way the British public have 'voted with their feet' and bought Japanese and German cars over the past 10 or 15 years in their search for reliability, as well as quality and after sales service. This has had the effect of bringing the British car industry practically to its knees and has greatly enhanced the prominence given to reliability and quality in the car industry. Clearly the reliability which is designed and built into a product is an important element of its marketing strategy. Here again, however, the importance of reliability must not be over-emphasised at the expense of other marketing considerations such as the competitiveness of pricing, the effectiveness of sales promotion and advertising, and the efficiency of after-sales service, as well as other product quality considerations.

Over the same time period the size, complexity, and sophistication of process plants have grown apace, with the capital costs increasing by orders of magnitude. The very fact that a piece of equipment such as a pump, costing a few thousand pounds, or one of its components, can cause a process plant costing millions of pounds to shut-down unexpectedly with loss of product and profits, and possible damage to other equipment, has forced the industrial purchasers of equipment to take a more sophisticated approach to purchasing decisions. This movement has seen: users of equipment becoming more critical of manufacturers claims of equipment performance; discussions and exchange of information on operating experience with other users before purchase decisions are taken; more information required in quotations and tighter guarantees; more detailed specification of requirements with reliability clauses included; and a more detailed critical analysis of the whole process of design used by the manufacturer/supplier. It has also seen the growth of user organizations, such as the Oil Companies Materials Organisation (OCMA) and the Engineering Equipment Users Association (EEUA), both of which have recently combined to form the Engineering Equipment and Material Users Association (EEMUA). These organizations were formed by user companies as a forum for an exchange of experience and information on equipment performance, as an attempt to derive common acceptable specifications, for lobbying manufacturers, and to provide a user input to the British Standards Institution deliberations. At the same time the larger companies in the process industries developed their own in-house expertise with the establishment of 'centres of excellence' in machines, vessels, piping, and, more recently, in the area of reliability. Organizations such as Shell, ICI, Courtaulds, BP, UKAEA, British Gas, and so on, all have reliability specialists whom other engineers can call upon to give advice, help analyse particular design problems, and assist with trouble-shooting on operating plants. Part of the task of these reliability specialists has been to assist in convincing sanctioning authorities that it is economic to increase capital expenditure, if this will improve reliability and availability, against strong pressures to keep capital expenditure down during a period of rapidly increasing inflation and capital costs.

(b) Safety

The world is becoming increasingly safety conscious. As regards human safety, the stringent design and maintenance specifications imposed on aircraft designers and operators by aviation regulatory authorities is an excellent reflection on the degree to which aviation engineers have managed to achieve high reliability in design and maintenance at a cost and level of risk which passengers seem willing to accept. To a large extent this has been achieved by very high standards of reliability in electronic control and navigational equipment coupled with the use of duplicate mechanical equipment. In reliability engineering terminology such duplication is known as 'redundancy', an aspect which is considered in detail in Chapter 7.

Safety considerations are also critically important where the integrity of industrial capital assets are concerned. This applies especially to plants which contain explosive and/or toxic or other dangerous substances, and those operating under severe operating conditions whereby their failure could result in loss of human life or injury and damage to plant and property, as well as economic losses to the owners. Prime examples of major industrial failures in recent times are the Chernobyl nuclear plant (USSR), the Bhopal chemical plant (India), and the Seveso chemical plant (Italy). Each of these failures has resulted in major loss of life, sterilization of large areas of countryside, far reaching international effects outside the immediate plant and country, and massive public reaction to the inherent dangers involved. In each case the failures have been the result of unreliability inherent in the design exacerbated by the human actions of the management and operators concerned. The economic and social consequences are almost beyond comprehension. These major hazards serve only to highlight the general safety problems that exist in the process industries. Different industries and processes plants have different degrees of risk depending on the nature of the product, the raw materials, and the processes involved. Clearly, the significance of reliability and its impact on safety will vary according to the degree of risk involved and the consequential effects of failure.

Many countries have enacted legislation which makes it mandatory for the manufacturers to ensure that no consequential harm to people occurs from the proper operation of the capital facilities and the proper use of the manufacturers products. In the United Kingdom the Health and Safety at Work Act (HASAWA) makes it a criminal offence if such harm does occur, and it places the onus of responsibility for the 'total safety' of the product firmly on the shoulders of the manufacturer. In addition a major hazards committee was established under HASAWA to control the design and operation of plants designated as major hazards. In the nuclear fields a similar role is carried out by the Nuclear Installations Inspectorate and public opinion is brought heavily to bear through Public Enquiries. All of this has had the effect of increasing the emphasis placed on reliability in its relation to safety.

(c) Product liability

Product liability is the legal liability of the manufacturer that the product is fit for the purpose for which it was sold, and includes consequential effects that arise if the product fails to meet this purpose. This factor is linked inextricably to the two previous factors of economics and safety. In its most simple form it could mean that the inability of a product to achieve its warranty specification results in the manufacturer bearing the entire cost of repairs or replacement. In other cases, however, the penalty to the manufacturer could be much more severe if the customer were indemnified by contract and/or in law against the consequential economic effects arising from the failure of such a product or equipment. The United States of America has seen some massive claims under 'product liability' legislation in the past few years. Although the legislative and legal situation in the United Kingdom is far from clear, experience with case law in the USA suggests that, with the passage of time, manufacturers can expect more litigation to arise from the failure of mechanical products. Equally important to the direct costs of such litigation is the damaging effect on company image and product sales from the adverse publicity which arises. Clearly, this further increases the pressure on manufacturers to improve the reliability of their products.

(d) Summary

Taking all three of these considerations into account there exist, in effect, good moral, legal, and economic pressures on companies to improve the reliability of their mechanical products and/or process plants, bearing in mind that such an objective must be tempered by the conflict of the needs and constraints which apply in particular situations and at particular times. This means that in the design of any mechanical product or any process plant system a *target* for *reliability* should be specified at the outset even though such a target may need to be modified later in the design process as a result of other considerations. Only in this way can the whole design team be made aware of the relevance and importance of reliability in the design process. In the USA, design specifications for reliability are now incorporated in procurement contracts for much of the equipment purchased by the Department of Defense, and evidence suggests that this need will diffuse rapidly into the private sector too. Under the aegis of EEMUA some large process industry companies in the UK have now started to include reliability clauses into equipment purchase specifications.

1.2 A definition of mechanical reliability

In common parlance there is a belief that a reliable product is 'good' and an unreliable product is 'bad'. Such subjective value judgments serve little scientific purpose, although they can be used, in a gamesmanship sense, in sales promotion and advertising. To foster a greater awareness and understanding of the problems associated with the study of mechanical reliability, however, we need as unambiguous and precise a definition of reliability as it is sensible to develop.

Several general definitions of reliability, covering all types of equipment, have been offered over the years, and most use similar words and effectively mean the same thing. A number of standards published by the British Standards Institution have included such definitions and include BS4778 and BS5760. For the special case of reliability for mechanical equipment the following definition is more appropriate and it has been adopted for the purpose of this publication:

Mechanical Reliability is the *probability* that a *component device or system* will perform its prescribed *duty* without *failure* for a given *time* when *operated correctly* in a *specified environment*.

NB

The key words, in italics, deserve special mention in order to promote a better awareness of the problems and intricacies involved with mechanical reliability.

(a) Probability

The concept of a probability was briefly mentioned in the introductory paragraphs. It is a concept which, in general terms, is alien to the understanding of most people, but is more likely to be better understood in terms of 'chance' or 'betting odds'. Its special significance is that it cannot always be predicted with certainty that some event will occur. Everyday occurrences relating to this involve weather forecasting, guessing the outcome of tossing a coin, *betting* which horse will win a particular race, and, more appropriately, *predicting* whether some equipment will fail to operate properly during the following week. Such events as these are known as random or *probabilistic* occurrences.

By contrast, other events, which are controlled by natural physical laws, can be predicted with certainty. Examples of these include the fact that night follows day, that an unpropelled object will eventually fall to the ground, and that, eventually, everyone will die. Such events as these are known as certain or *deterministic* occurrences.

In effect, therefore, the mathematics of probability is the mathematics of uncertainty, in that it is not possible to explain precisely the combinations of physical, and sometimes social, laws which affect the outcome of everyday chance occurrences. It is the mathematics of analysing chance events and predicting the likelihood of events occurring during a given period of time. Although the mathematics of probability are well developed in the literature, it is the intention here to keep to simple probabilistic concepts and notation which are sufficient for a basic understanding for the application of methodology and techniques. Readers who are more deeply interested in the theory and require a better understanding of the underlying mathematics are advised to examine the reading list provided in the References at the end of this book.

Probability can be considered as the chance of some event happening, such as a pump failing to operate because its drive motor has burnt out. For example, if it is assumed that there are 100 machines on test initially, the numbers that fail during specified time intervals (say, each week) could be counted and the results illustrated with the use of a histogram. Such a histogram is shown in Fig. 1.1, where the vertical ordinate shows the number of machines which fail in each time period, and time is measured in weeks. We can make the following observations from this histogram.

(1) No machines failed during the first period from 0 to 10 weeks.

(2) All machines had failed by 70 weeks.

(3) Fifteen machines failed during the first 30 weeks.

(4) Seventy-five machines failed during the period (30–60) weeks.

In predicting the probability of failure of a batch of similar machines, operating under similar conditions, it could be predicted that in:

(1) 0–10 weeks	zero failures	Reliability $R = 1$
(2) 70+ weeks	100% failures	Reliability $R = 0$
(3) 0–30 weeks	15% failures	Reliability $R = 0.85$
(4) 30–60 weeks	75% of items *not* failed by 30 weeks are likely to fail by week 60. This is a conditional probability – the probability that an item will fail in the period 30–60 weeks *given that it has survived the first 30 weeks*. At this stage conditional probability will not be considered further (see section 15.2(b)).	

To obtain a better understanding of the mathematical language and notations used to describe mechanical reliability, this histogram needs to be modified to give a measure of the 'relative frequency of failure', rather than the numbers of machine failures, during each time interval. The 'relative frequency of failure' is the number of machines failing, $n(t)$, during a time interval, divided by the total number, N, on test (in this example, 100). This is shown in Fig. 1.2 and gives an almost identical histogram to Fig. 1.1, except that the ordinate values are all less than unity.

The histogram in Fig. 1.2 leads directly to the definition of the probability of a *single machine* failing between an upper time limit t_2 and a lower time t_1. This is given by the equation

Probability of single machine failing
$$= F(t_2,t_1) = \{n(t_2,t_1)\}/N \quad \text{between } t_1 \text{ and } t_2$$

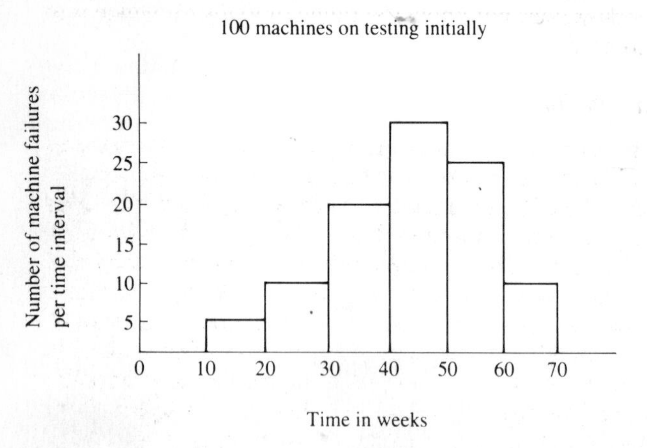

Fig. 1.1. A histogram of the failures of machines

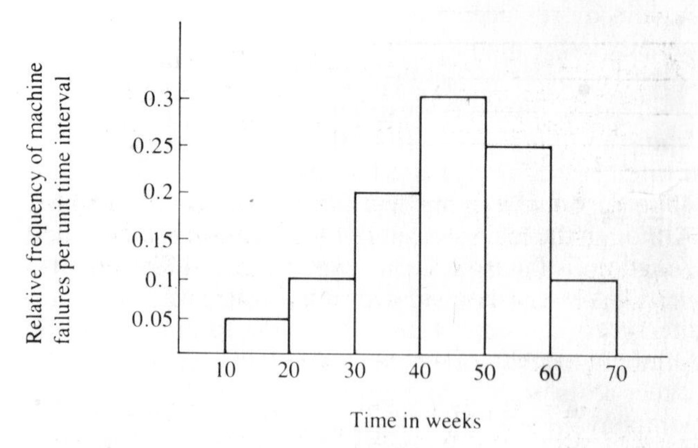

Fig. 1.2. Histories of the relative frequency of failures of the machines

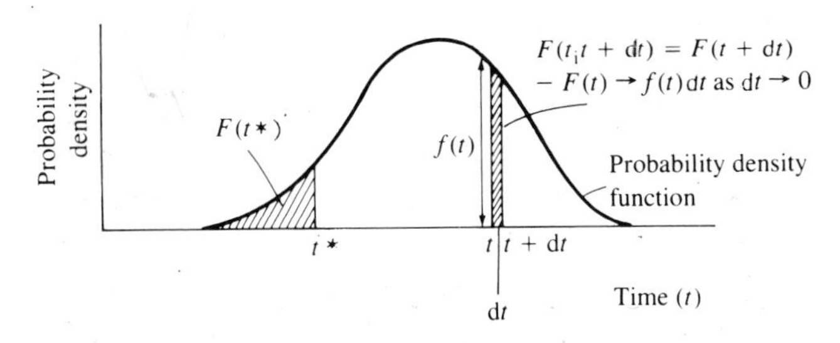

Fig. 1.3. Probability Density Function (pdf) for the failure of a machine population

where

(1) $n(t_2 t_1)$ denotes the number of failures occurring in the time interval t_1 to t_2;
(2) N denotes the number of machines in the sample;
(3) the probability of failure $F(t)$ is measured as a decimal (rather than as a percentage).

If the time intervals of the histogram were reduced until they became infinitesimally small (dt) a smooth shaped distribution of failures curve, similar to that shown in Fig. 1.3, would be obtained. The term 'relative frequency' could then be replaced by the term 'Probability Density Function' (often referred to as the pdf) which is fundamental to the mathematics of reliability. Because the Probability Density Function is a function of time, it is represented by the notation $f(t)$.

In this case the probability of a single machine failing during the time increment dt (that is, $t, t + dt$) is given by

$$F(t, t + dt) = f(t)\, dt$$

which is shown by the shaded column on the right of Fig. 1.3, and the probability of a single machine failing between 0 and the time t^*, which is denoted by $F(t^*)$, is shown by the shaded area to the left of Fig. 1.3.

Since the probability of an item failing up to a given time is complementary, in a mathematical sense, to the probability of the same item's survival (that is, its reliability) it follows that

$$F(t) + R(t) = 1 \quad (1.1)$$

where

(1) $F(t)$ is the probability of failure before time t;
(2) $R(t)$ is the probability of survival to time t or the reliability of the single item (component, device, or system).

The probabilistic concepts of mechanical reliability will be considered in more detail in Chapter 2, but the significance of other key words in the definition of mechanical reliability must now be considered.

(b) Component, device, or system

A component is considered the smallest part in an assembly which would normally be replaced by a new component, but sometimes repaired, on failure. Typical mechanical examples would be springs, bolts, and impellers. By contrast, a device would comprise several, possibly many, components, and typical examples would be pumps, compressors, and gear-boxes. Lastly, a system would comprise several, possibly many, devices, and in this case typical examples would involve aircraft, motor-cars, factory production and packaging lines, and process plant systems.

The detailed mathematics of mechanical reliability can be quite different for components, devices, and systems, as will be shown in Parts Two and Three, so it is very important to state precisely the physical boundaries of the item whose reliability is under review.

(c) Duty

The more severely an item is stressed the greater is its probability of an early failure. For example, a standard motor-car which is driven to the limits of its capability is likely to fail sooner than one which is driven at a modest speed. In effect, therefore, the service or 'duty' which is expected of an artefact is of prime importance to its reliability specification, because this describes the stresses which will arise during its *normal operation*. Herein lies one of the greatest differences which distinguishes mechanical from electronic reliability. Normally most electronic equipment is operated in a benign way, whereas mechanical equipment can be subjected to wide ranges of operating stress. Consider, for example, the case of a centrifugal pump.

During a start-up operation, it may consume more than three times its normal operating power to overcome inertia, during which its components are subjected to high start-up stresses. Similarly, such a pump can be operated in a variety of manners ranging between 'fully throttled' to 'fully open', with considerable differences in pressure-related stresses.

It will be appreciated, therefore, that the design for mechanical reliability will normally require a far greater analysis of the likely operating stresses of an item than is required for electronic equipment.

This highlights another problem with mechanical equipment where, particularly with standard off-the-shelf equipment, the purchaser does not clearly know the range of stresses for which it was designed, and the supplier does not know the range of loads for which it is required.

(d) Failure

Another aspect which often differentiates mechanical and electronic reliability is the definition of what constitutes the failure of an item to perform its duty. Whereas an electronic component may fail to render its duty because of a simple open or closed circuit breakdown, it is often more difficult to prescribe those conditions which constitute the failure of mechanical equipment which might be able to perform all or part of its duty in an impaired state. For example, a process valve might be retained in operation despite its leaking. The question of what constitutes a product's failure to fulfil its duty, therefore, is a moot point which must be decided by the people who are most informed and concerned with the consequential effects of that failure. A valve leaking water might not be classified a failure until a very substantial leak occurs. If, however, the leaking fluid were highly corrosive, toxic, or inflammable, then even the smallest leak might not be tolerated, and thus classified as a failure. Normally, the failure of electronic equipment does not incur this process fluid consequential effect.

To improve the classification of failures it is normal practice to relate to their *modes, causes*, and *effects*.

A failure *mode* is the way in which that failure is made manifest. For process valves, for example, the modes of failure could be:

(1) leaking internally – failing to seal when in closed position;

(2) leaking externally – open or closed position;

(3) inoperative in the open position – will not close on demand;

(4) inoperable in the shut position – will not open on demand;

(5) spurious operation.

The *causes* which brought about such failures could be examined at different levels of detail. For example, a crude analysis might suggest that improper adjustment of the valve gland, degraded material, deposits, or substandard components led to the failure. A more exact analysis might suggest that corrosion, erosion, distortion, and fatigue, as well as maloperation, were the root causes of failure.

To qualify as a failure, however, its *effect* must be such as to prevent the device from sustaining its prescribed duty.

Failure causes, modes, and effects will be covered in more detail in Chapter 14.

(e) Time

It should be emphasised that the variable, or combination of variables, which affect materially the failure of an item should ideally be used to define its reliability. Sometimes the number of cycles which the item performs may be the governing factor which affects its failures. Examples include the number of on-off operations of a switch, or the number of machine starts or its number of revolutions, to failure. Such examples as these, however, necessitate the use of monitoring devices, which involve the use of resources and, in themselves, create reliability problems. In practice, therefore, it is usual for reliability engineers to rely on the elapsed time in the operation of a machine before it fails as the variable of interest, albeit that time might not be the most accurate variable to use. It is for this reason alone that time was used as the variable in section 1.2 (a).

(f) Operated correctly

It will be appreciated that the misuse or, indeed, the abuse of a piece of equipment can adversely affect its reliability. It is, therefore, incumbent on the designer and the product warranty manager to specify the limits within which the equipment should be used. This requires a proper understanding of the customer's requirements and skills and necessitates high standards in the design of operating and maintenance manuals. Also, it would be folly to introduce highly sophisticated machinery into a plant if the existing maintenance staff comprised rough and ready pipefitters, without making provision for the recruitment and training of appropriate maintenance staff.

(g) Specified environment

It follows that the more severe the stress imposed on a piece of equipment by its environment the more likely it is to fail.

Here again lies one of the greatest differences which distinguishes mechanical from electronic reliability. Normally electronic equipment is operated in as benign an environment as possible, which in some cases will involve its complete encapsulation to avoid contamination, corrosion, erosion, and fatigue. By contrast, mechanical equipment is often expected to operate between environmental extremes, such as a motor-car working in both desert and arctic conditions. Again it behoves the designer to cater for such extremes at the outset of his design and, of course, this necessitates good market research or close liaison with the client.

At this stage it will be appreciated that, while there may be deficiencies with this definition of mechanical reliability, because the subject is complex, it can nevertheless prove helpful, provided readers are aware of the many variables involved and their limitations.

CHAPTER 2. THE CONCEPT OF MECHANICAL RELIABILITY

To develop the understanding of mechanical reliability to a point where its application can be practised two further aspects of the subject need to be considered. These are the use of interaction diagrams, and the bath-tub curve.

2.1 Interference diagrams

If it is assumed that a component has some distribution of strength designed into it and another distribution of load (stress) imposed upon it, then it should be clear that if these probability density functions (pdfs) are completely separated (that is, there is no interference between them), the result in theory, is an absolutely reliable component. This is illustrated in Fig. 2.1, where

$\bar{S}$ = mean value of the component's strength

$\bar{L}$ = mean value of the load imposed on the component

σ_S = standard deviation of the component's strength

σ_L = standard deviation of the loads imposed.

Standard deviation is a simple (statistical) measure of the variation of some attribute (in this case strength or load) about its mean value

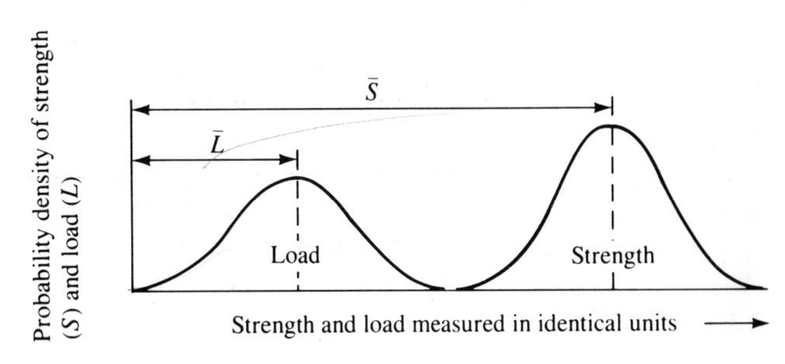

Fig. 2.1. Strength – load interference diagram with complete separation

If, however, the Probability Density Function curves for both strength and load are shown to interfere, in that they intersect one another, as shown in Fig. 2.2, then in this case the shaded area where the curves intersect is an indication that the component will fail, because the load exceeds the strength.

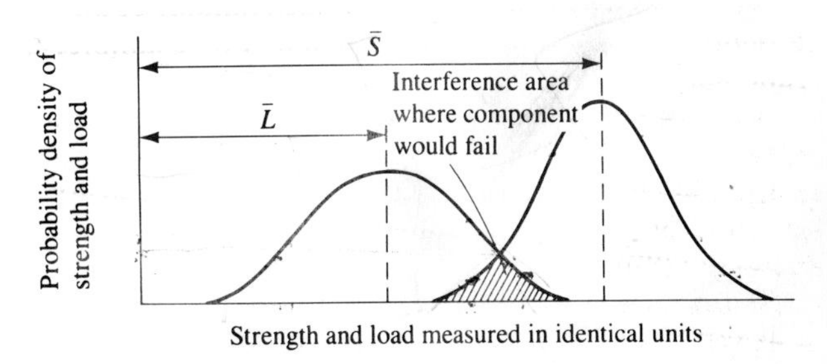

Fig. 2.2. Strength – load interference diagram with interference

An established technique, related to strength and load interference, used in design, is the use of safety margins. When designing reliability into a component the term 'safety margin' (SM) takes on a special meaning and is defined by the following relationship

$$SM = \frac{\bar{S} - \bar{L}}{\sqrt{(\sigma_S^2 + \sigma_L^2)}} \qquad (2.1)$$

For normally distributed load and strength, it can be shown that the relationship between a component's reliability and its safety margin is given by the graph shown in Fig. 2.3.

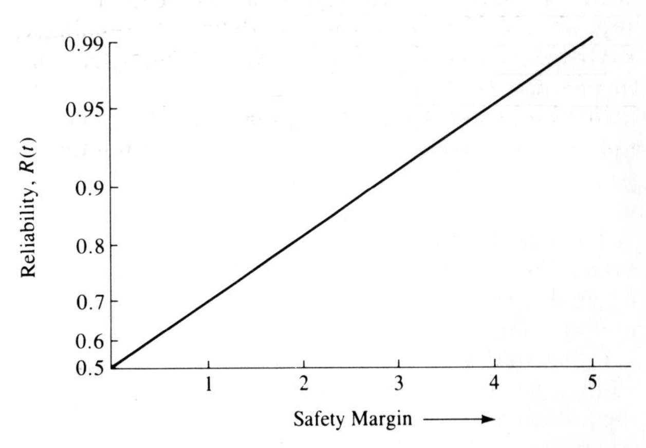

Fig. 2.3. Relationship between a component's reliability and its safety margin

Variations in the strength of the component may result from a combination of variations in the metallurgical and fabrication processes involved with its manufacture, whereas variations in the load imposed upon it result from variations in its duty and environment. Over the years, materials scientists and mechanical engineers have invested much time and great resources in studying and improving our knowledge of these variations, but currently the variation of the load imposed on many mechanical components (σ_L) is the most poorly understood variable affecting a component's reliability.

A basic assumption which is made when Interference Diagrams are used is that the probability density function of a component's strength remains static with repeated load applications. This is often true with electronic components, where the safety margin technique can be used profitably. However, for mechanical components, whose strength often deteriorates with successive application of duty due to wear, fatigue, and corrosion, the assumption is often unfounded. This implies that, with use, the Probability Density Function of strength for a mechanical component would move to the left of the Interference Diagram of Fig. 2.2. As a consequence this would increase the area of interference, increase its chance of failure, and reduce its reliability. Similarly, if the major loads (stresses) imposed on that component were derived from another component, which itself was also deteriorating and causing an increase in load, then the Probability Density Function of load for the first component would move to the right of the Interaction Diagram of Fig. 2.2, with a consequential reduction in its reliability.

A more detailed understanding of the use of Interference Diagrams and the effects of strength and load changes for mechanical components can be obtained by reading the work of Carter (**1**)*.

2.2 The Bath-tub Curve

A fundamental concept in reliability engineering is the Hazard Rate Function, $z(t)$. Other terms used for this function are:

force of mortality;

age specific failure rate;

instantaneous failure rate;

conditional failure rate;

hazard function.

These are often incorrectly referred to as the 'Failure Rate' in reliability circles. It is essential to recognise the difference between the Hazard Rate Function and the Failure Rate as they will, except in special cases, have different numerical values, and this can cause great confusion in problem-solving situations. The term Hazard Rate Function is used to describe the behaviour of *non-repairable* components which form part of a system. The term *Failure Rate* implicitly assumes that the time to failure distribution is exponential and is used to describe the behaviour of *repairable* systems. This aspect is covered in greater detail in Part Two, but for the purposes of discussion in Part One, where the more general case is dealt with, the term hazard rate function will be used.

The Hazard Rate Function, then, is a measure of the probability that a component will fail in the next time interval, given that it has survived up to the beginning of that time interval.

Using the same notation which has been used previously, it can be proved that this probability is given by the following

* References are given at the end of the book.

Probability of failure in next time interval

$$= z(t)\ \mathrm{d}t = \frac{F(t + \mathrm{d}t) - F(t)}{R(t)}$$

that is, the probability of failure at time t, and given that no failure has occurred before time t, which can be simplified as follows

$$z(t)\ \mathrm{d}t = \frac{f(t)\ \mathrm{d}t}{R(t)}$$

so that the Hazard Rate Function

$$z(t) = \frac{f(t)}{R(t)} = \frac{\text{Probability density function}}{\text{Reliability}} \qquad (2.2)$$

A method for calculating the Hazard Rate Function is given in Chapter 6.

A special reason for introducing this concept is that it is commonly assumed to exhibit the profile shown in Fig. 2.4, which is known as the 'Bath-tub Curve'.

The three phases of the 'Bath-tub Curve' are described below.

(a) Phase 1

Phase 1 shows a declining $z(t)$ due to the premature failure of components – similar, for example, to genetic deficiencies in some babies; for this reason this phase is often called the 'infant mortality phase'.

(b) Phase 2

Phase 2 shows an approximately constant $z(t)$ due to the chance failure of the component because of its unexpected over stressing. An example of this phenomenon would be the 'accidental dropping and breaking of a plate'. Sometimes failures which occur in this phase are called 'random failures'. This phase is known as the component's 'normal operating life'.

(c) Phase 3

Phase 3 shows an increasing $z(t)$, meaning that the probability of a component failing between equal and successive time intervals *increases*. This is similar to the phenomenon of 'ageing' in people after the first 25 to 30 years and gives rise to the title 'wear-out phase'.

While this description is conceptually useful it should not be taken too literally. One reason for this reservation

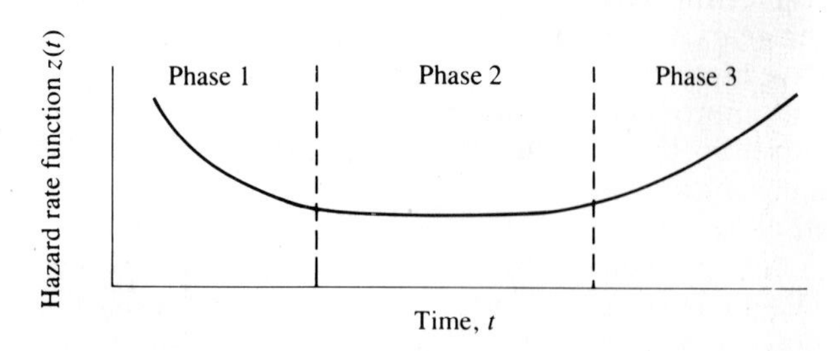

Fig. 2.4. The conceptual 'Bath-tub' Curve

is that all three types of failure regime can occur simultaneously, but with varying degrees of severity, *so a graph of* z(t) *is specific to the type of component and the way in which it is stressed.*

Such a graph could be considered the component's '*survival signature*'; that is, an inbuilt characteristic of the component which determines its likelihood of survival in specified circumstances.

This leads naturally to the subject of reliability life testing. Normal commercial practice requires that most electronic equipment should be tested at stresses somewhat higher than normal rating with the purpose of failing those weak components, attributed to Phase 1, before their despatch to customers. Of course, too high a stress must be avoided lest more than the proportion of failures ascribed to Phase 1 should occur. In electronic engineering parlance, this procedure is called 'burn-in'. As a consequence, the values of $z(t)$ for many electronic components in active service are fairly constant over large intervals of time, and it is this phenomenon which has led many people to *assume*, *incorrectly*, that the same holds true for mechanical components. Unfortunately, this misconception also results from the erroneous analysis of mechanical component failures, a factor to which we shall return later. Instead a thorough analysis of the reliability of mechanical components shows that many exhibit a 'wear-out' tendency, with increasing values of $z(t)$ throughout their operating lives, and it is for this reason that mechanical equipment is given a whole range of maintenance attention to reduce the effects of failure in a cost effective way.

Clearly, reliability life testing is an important element in reliability engineering design, and this highlights another big difference between electronic and mechanical reliability. Frequently electronic equipment is sufficiently inexpensive and small that large random samples can be tested to failure, relatively cheaply, under controlled conditions of duty and environment, thereby providing the excellent reliability data and accurate values for $z(t)$, which have been published widely. By contrast, however, mechanical products tend to be much more expensive to purchase and operate, so the tendency is for mechanical *components* only to be tested by these means. It is rare for large samples comprising complete mechanical products to be tested to failure. Instead, their reliability attributes tend to be measured by their in-service performance, which involves many problems associated with the homogeneity of the sample and the similarity of their operating and environmental regime. Furthermore, because of the rate of innovation in electronics, such equipment tends to be replaced due to obsolescence before the wear-out phase is ever reached, if indeed it exists for such equipment.

CHAPTER 3. OPERATIONAL AND COST IMPLICATIONS

3.1 The effects of maintenance policy on in-service reliability

The failure of equipment in service can be mitigated by a variety of maintenance policies whose rationale depends, to a large extent, on the failure characteristics of the items – that is, decreasing, constant, or increasing Hazard Rate Function.

Two major maintenance policies and the effect of these different failure regimes are discussed below.

(a) Preventive maintenance

This is the policy whereby some inspection or rectification of the component, device, or system is carried out prior to its failure, based upon its condition, the elapsed time it has been in service, or its exposure to stress. This policy also includes inspection or rectification when the opportunity arises due to plant shutdown caused by other component failures or for operational reasons. Routine maintenance actions such as topping-up oil in gear boxes, lubrication, condition monitoring, and so on, carried out without taking the equipment out of service, are also included.

For those circumstances which necessitate taking a component out of service for preventive maintenance with a view to returning it to its 'as good as new' (AGAN) condition, in a reliability sense, it can be proved that such a policy is ony justified if two conditions prevail, namely:

(1) the failure regime should exhibit a significantly increasing $z(t)$;

(2) it can be shown that such preventative maintenance is also justified in either an economic, safety, or moral sense, or any combination of these.

(b) Corrective maintenance

This is the policy whereby the repair or replacement of a component only takes place after the component has failed in service.

By corollary with the statement for preventive maintenance, it can be shown that corrective maintenance can be justified if the Hazard Rate Function, $z(t)$, is approximately constant. In this case there is *no justification* for taking the component out of service for preventive maintenance since, at best, its probability of failure would remain unchanged or, at worst, it could deteriorate.

3.2 The 'AGAN' philosophy

The 'AGAN' philosophy has been explained for three basic reasons:

(1) to help the reader understand the logic of the preceding material;

(2) to introduce the idea of the reliability of equipment deteriorating with successive repairs;

(3) to introduce the notion of reliability improvement due to the phenomenon of 'learning'.

Research has shown that, after repair, mechanical equipment is not 'as good as new' in a reliability sense. A few examples will explain this. If the example of a shaft supported by several bearings which exhibit 'wear-out effects' (that is, the Hazard Rate Function, $z(t)$, is increasing) is considered, it will be appreciated that the failure and replacement of one bearing by another with the same reliability characteristics (that is, survival signature) cannot return it to its 'AGAN' condition in a reliability sense. Even if the shaft system were able to resume its full functional duty it would not be 'AGAN' because the other bearings would have attained higher values of the Hazard Rate Function, $z(t)$, and would be likely to fail in a shorter time interval. This same phenomenon can exist even when devices or systems receive a thorough overhaul or reconditioning, because the bodies or structures which provide the framework for the device or system often exhibit 'wear-out effects' themselves, due to distortion and creep. In addition, if equipment is repaired locally, maintenance and reassembly standards are rarely as good as those prevailing in the original factory. As a result maintenance errors or poor quality workmanship (often influenced by the environment) jeopardise the integrity of the device and prevent it from returning to an 'as good as new' condition from a reliability point of view.

As a result of phenomena such as distortion, creep, and the quality of maintenance, several researchers have concluded that after repair the reliability of much mechanical equipment is 'as bad as old' (ABAO). This means that the Hazard Rate Function, $z(t)$, immediately after repair approximates to its value just prior to failure (that is, although the device may function satisfactorily after repair its risk of failure is no better than before the repair). Clearly it would be incorrect to assess the reliability of mechanical equipment without properly considering the effects of maintenance, and it could be fundamentally misleading to attempt to plot the values of the Hazard Rate Function, $z(t)$, for a sample of mechanical equipment subjected to different repair regimes, on the same graph. A detailed reading of the literature on reliability engineering shows that most authors have ignored this phenomenon. A method for dealing with this problem is given in Chapter 6.

Maintenance engineers will generally endeavour to improve the reliability of equipment with successive repairs. This is usually possible during the early operating experience with new designs and where there is successive development of new models which attempt to design out earlier failures due to design, operating procedures, or maintenance deficiencies. In some cases, therefore, it can be seen that the reliability of some equipment can improve through 'learning', such that the values of the Hazard Rate Function, $z(t)$, decrease with successive maintenance actions. Duane (**2**) produced a learning-curve model to study these learning effects.

The upshot of both these considerations of reliability deterioration and reliability improvement (through 'learning') is that the 'Mean Time To Failure' (MTTF) for a sample of *mechanical* devices can decline, rise, or oscillate, depending on which of these effects is the stronger. This can depend as much on the quality of maintenance management and service as it does on any intrinsic design or reliability attributes of the new device or its spare parts.

Conversely, although the reliability of *electronic* components, devices, and systems can be enhanced in a similar way, through learning, most electronic equipment does *not* suffer deterioration in reliability from maintenance action. This is partly because electronic equipment maintenance is very largely a matter of simple circuit board replacement, and the intrinsic reliability of electronic devices does not rely (generally speaking) on the effects of inter-connected electronic equipment as it does with mechanical equipment. Therefore the strong failure knock-on or cascade effects between components is avoided.

3.3 Availability

Besides endeavouring to maintain the functional duty of equipment, the purpose of maintenance is to optimise its in-service life. This involves three factors:

(1) increasing the time to failure;
(2) decreasing the downtime due to repairs and maintenance;
(3) achieving (1) and (2) in the most cost-effective manner.

Such considerations as these give rise to another attribute of reliability engineering: Availability. This, in simple terms, is the proportion of time that a component, equipment, or system is capable of performing its duty. A measure of availability is given by

A = Availability (of device equipment or system)

$$= \frac{\text{MTTF}}{\text{MTTF} + \text{MTTR}} \tag{3.1}$$

where

MTTF = Mean time to failure

(Note that this may equal the reciprocal of the hazard rate function, $z(t)$, but only when this is constant)

and

MTTR = Mean time to repair

Clearly, the greater a machine's availability, the greater its capacity to earn its keep will be, and the lower are the consequential losses which result from its failure and subsequent maintenance.

A maintenance manager's role is more than to improve the reliability of the plant and equipment in his charge – it is to improve its availability in a cost-effective manner. This can be achieved either by improving its reliability (MTTF) or by reducing the time taken to carry out repairs (MTTR) or a combination of both.

3.4 Life Cycle Cost implication

In the same way that there is a human life-cycle of birth, growth and development, old-age, and eventual death, so there is a 'life-cycle' for mechanical equipment of design and prototype development, manufacture, in-service development, maintenance, obsolescence, withdrawal from service, and eventual scrapping or disposal. The costs associated with a piece of mechanical equipment through the whole of this 'life-cycle' are known as its Life Cycle Costs (LCC). This was touched upon in Chapter 1 and here it is intended to expand on the implications that mechanical reliability has for life cycle costs. These costs include:

Design costs;
Technical and market research and development costs;
Prototype fabrication and testing costs;
Production costs;
Warranty costs;
Equipment capital costs;
Working capital costs;
Operating costs;
Repairs and maintenance costs;
Downtime or lost opportunity costs;
Replacement costs;
Retirement and disposal costs.

All of these costs are influenced in some degree by the reliability of the equipment.

The best opportunity for minimizing the Life Cycle Costs occurs during the design stage before decisions are made which preclude consideration of further competing design alternatives. Ideally the designer should select that level of reliability which, in concert with all the other attributes specified for a product, will minimize its life cycle costs. This is easy to specify but more difficult to achieve in practice without long experience in the cautious development of products which gain general acceptance. Clearly, for those industries working in the forefront of technology, greater risks need to be taken, with the possible consequence of higher Life Cycle Costs.

Although material scientists and mechanical engineers have contributed handsomely to the understanding of the strength of materials, the analysis of stresses and strains, and the success/failure attributes of certain designs, it is clear that the technology of mechanical reliability needs to be developed much further before it is possible to optimize the Life Cycle Costs of a brand new product. Two prerequisites to achieving this end are as follows.

(1) Adequate recording and analysis of in-service reliability data.

(2) The feedback of all performance and reliability data to the designers in order to influence:
   (i) design modifications for existing products;
   (ii) design improvements for new generation products;
   (iii) development of relevant and cost effective data banks which form an essential link in the search for engineering excellence.

In the search for this optimum design, trade-offs are reached between the various costs. For example, it would be reasonable to expect capital cost to increase with improved reliability (but not necessarily so). This would arise because design time, more testing, improved materials and methods of fabrication, better data collection and retrieval systems, and so on, would be needed in the search for improved reliability. Furthermore, it would also seem reasonable to expect the law of diminishing returns to apply to reliability improvements above a certain threshold, such that capital costs would increase faster than the gains from improvements in reliability. Therefore, the capital cost versus reliability relationship would resemble the graph shown in Fig. 3.1. By contrast many other costs would be expected to reduce with increasing reliability. Improved reliability should result in lower downtime, increased availability, lower maintenance costs (including spares consumption), and greater earning capacity, as illustrated by the graph in Fig. 3.1. The addition of the increasing capital costs curve to the reducing curve for other Life Cycle Costs would give a hypothetical curve for the total Life Cycle Costs with an optimum point, at *x*, to give a minimum Life Cycle Cost.

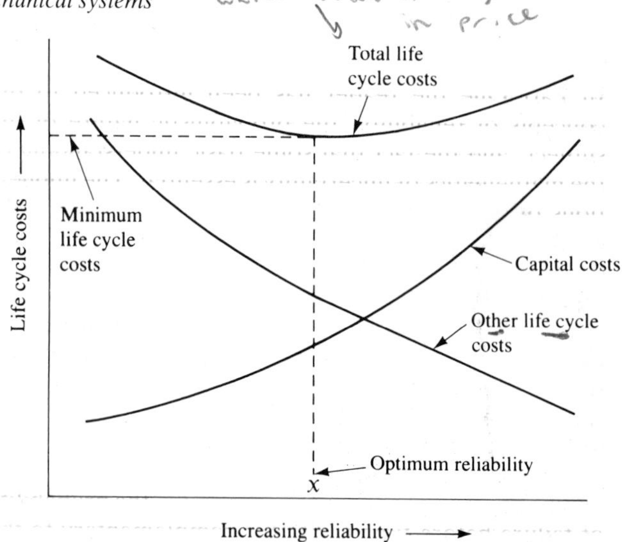

Fig. 3.1. Conceptual model of the life cycle costs

Although the model shown in Fig. 3.1 is useful in a conceptual sense, there is no published data to demonstrate that this has been achieved, in practice, for any particular product. However, published data does exist for certain parts of this optimization process. For example, Jardine (**3**) has shown how to minimize the cost of operating a large fleet of vehicles, and de la Mare (**4**) has demonstrated how the Mean Time To Failure (MTTF) for process equipment deteriorates with each successive repair, and that optimal replacement strategies exist which could realise substantial economies in the Life Cycle Costs.

CHAPTER 4. SUMMARY

In Part One the reader has been introduced to the concepts of reliability, the probabilistic nature of the subject, and the fact that some basic understanding of the mathematics of uncertainty is essential. The implications of mechanical reliability for product economics, product safety, and product liability were then discussed in order to provide a background to the more technical aspects of the subject.

The reader was then introduced to a definition of mechanical reliability, and the precise meaning and significance of the key words in this definition were spelled out. Simple mathematics and graphics were used to bring out the meaning of the key concepts of the probability density function ($f(t)$ and the probability of failure before time t, ($F(t)$). The fact that the probability of failure before time t ($F(t)$) is complementary to the probability of survival to time t, that is, reliability, ($R(t)$), was spelled out, and that, *mathematically, the probability of failure before time* t *plus the probability of survival to time* t *equals unity*.

That is, $F(t) + R(t) = 1$

The key concepts of mechanical reliability were further explained with the use of interference Diagrams and the Bath-tub Curve, with its three phases of 'infant mortality', 'normal operating life' and 'wear-out'. This introduced the concept of the 'Hazard Rate Function', $z(t)$, and how this behaves during the three phases of the Bath-tub Curve. The essential differences between mechanical and electronic components were explained, and these are summarized in Table 4.1.

The concept of the 'as good as new' philosophy was introduced and again the difference between electronic and mechanical components and equipment in this respect were drawn out. For mechanical equipment the conflicting notions of deterioration in reliability with successive repairs and improvements in reliability through 'learning' were explained. Further, that while a constant Hazard Rate Function, $z(t)$, for electronic equipment can generally be assumed, for mechanical equipment it can either decline, rise, or oscillate depending on which of the conflicting effects is stronger.

The concept of Availability was introduced and its relationship to the two other factors of the 'Mean Time To Failure' (MTTF) and the 'Mean Time To Repair' (MTTR) was discussed in order to establish that

$$\text{Availability } (A) = \frac{\text{MTTF}}{\text{MTTR} + \text{MTTF}}$$

and that a maintenance manager can improve plant availability by either increasing its reliability (MTTF), by reducing the time taken by repairs (MTTR), or by a combination of both.

Finally, the concept of Life Cycle Costs was introduced, and the implications this has for mechanical reliability were demonstrated by the use of a simple conceptual model.

In Part Two the reader will be introduced to the analysis of reliability data collected during in-service operating experience, which provides the basis for reliability assessment of new process plant design.

Table 4.1. Comparison of electronic and mechanical equipment

Attribute	*Electronic equipment*	*Mechanical equipment*
(1) Modes of failure	Simple	Complex
(2) Stress factors	Predictable	Difficult to predict accurately
(3) Burn-in to remove failures	Can be economically justified	Normally too expensive
(4) Constant hazard rate	Normally applies for long duration	Applies for short duration and often not at all
(5) Increasing hazard rate	Early obsolescence usually precludes this phase if, indeed, it ever exists	Usually occurs very early in operating life, often from the onset
(6) Life expectation	Short due to obsolescence	Long
(7) Life testing	Cheap and effective	Difficult and expensive
(8) Preferred maintenance	Replace rather than repair	Repair and replace
(9) Reliability data	Well documented	Very little good data

PART TWO

Analysis of in-service reliability experience

CHAPTER 5. INTRODUCTION

Part One has introduced the principles, concepts, and philosophy of reliability engineering, stressing the probabilistic nature of reliability, and emphasising the differences between electronic and mechanical reliability. Part Two will now look at the basic techniques available for the analysis of in-service reliability experience.

At this stage it is important to stress that engineering is an applied science (not a pure science), operating in the real world with imperfect materials, and many unknown parameters (that is, imperfect information). The greater level of uncertainty involved with mechanical equipment has already been stressed in Part One and yet, as the size and complexity of process plants increase, with increases in the size and duty of mechanical equipment beyond established experience of manufacturers and users alike, plant managers and engineers are increasingly being called upon to ensure efficient and reliable operation.

The prediction of how the equipment he is responsible for will behave in operation is a fundamental part of an engineers role. Design engineers must predict how equipment or systems they have designed will operate and perform some specified duty in months or even years to come when the plant is commissioned, and plant/maintenance engineers must predict how key equipment or systems will perform after each overhaul, shut-down, or major modification. Sadly, this is largely carried out on the basis of overall experienced 'judgement', without the benefit of formal analysis of past experience, and too often design engineers are isolated from any feedback of operating experience.

As it is almost impossible to conceive, within the limits of commercial constraints, that a large, multi-million-pound process plant would operate perfectly with high reliability from the word go, the real process of predicting reliability and bringing process plants up to high levels of reliability is an iterative one. Thus, the reliability behaviour of process plants will depend on good initial plant design; well designed systems for the recording of plant operational experience and the analysis of this recorded experience; cost effective plant improvements; further analysis of post-improvement operating experience; possible further improvements, and so on.

A simple model of this iterative process of bringing plants up to a cost-effective level of reliable operation is shown in Fig. 5.1.

A key vital part of this iterative process is the operation of efficient and effective feed-back loops (shown by broken lines in Fig. 5.1) which consists of the collection of operating data, the formal analysis of this data to meet a specified need (that is, reliability prediction), and the feedback of the results to the appropriate decision-making centres in the organization. Too often it is found that expensive systems for collecting data are established without thinking through what specific needs the data will meet, and without setting up any formal system for analysis.

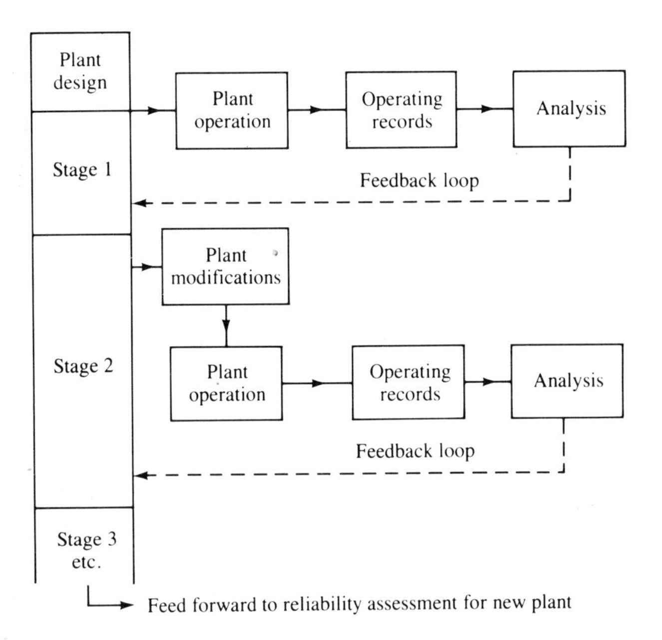

Fig. 5.1. Reliability improvement model for process plant

The problems of data collection will be covered in Part Four. As a pre-cursor to the reliability assessment of new plant design, which is dealt with in Parts Three and Four, the basic techniques available for the analysis of in-service experience is dealt with here in Part Two. Chapter 6 will cover the analysis of operating experience for mechanical components, and Chapter 7 will cover the analysis of operating experience for mechanical repairable systems.

A final point here is that all the clever analysis of past operating experience is of no avail unless the information resulting from the analysis is passed on to those decision centres where the information will be used for the prediction of how equipment or systems are likely to behave in the future.

CHAPTER 6. ANALYSIS OF IN-SERVICE EXPERIENCE FOR MECHANICAL COMPONENTS

6.1 Introduction

The analysis of reliability experience for mechanical components, which form parts of mechanical repairable systems, is a vital part of reliability engineering.

To clarify the distinction, it is convenient to define a *component* as an item which can only fail once (*cf.* definition in section 1.2(b)). If such a failed component forms part of a repairable system, the system is repaired by *replacement* of the failed component.

In turn, *replacement* can be viewed as the restoration of the 'hole' in the system (brought on by the component failure) back to the 'as good as new' (AGAN) condition. This can be achieved either by a physical replacement of the item by a new one of the same type, or by perfect repair of the 'hole'.

This assumption of as good as new replacement is a very important one in component reliability analysis. Such analysis includes an assumption that all observations are of components of the same type (in a statistical sense) – that is, the lifetimes are independent and identically distributed (IID).

Component reliability analysis, as described in this section, is based on the observation of times to failure (and to non-failure) of components in service, or in testing. The information obtained will be useful for the following

- identification of problem areas;
- aiding engineering investigation of causes of failure;
- specifying planned maintenance and replacement strategies;
- quantifying spares requirements;
- reliability assessment for new plant design.

6.2 Lifetime distributions

The measure of the reliability of an individual component is its 'lifetime' – the 'time' elapsing between its start of life and the 'time' at which it fails.

The quotation marks around 'time' are intended to convey the idea that, whilst it is usual to talk in terms of that variable, and use the symbol t, it does not necessarily imply the passage of 'clock time'. It represents any suitable measure of component usage, and it is a matter of engineering judgement to choose the right one.

The symbol t might, quite often, represent a straightforward elapsed time. Alternatively, it could represent any of the following.

Operating time (for an application where time during which the component does not operate does not include any failure inducing aspects).

Operating plus 'stand-by' time.

Distance covered (many motor vehicle component failures are km-dependent rather than age-dependent).

Number of missions (for example, in aircraft, the number of flights is usually the most relevant variable).

Throughput volume (of chemicals, and so on, or of gases in petro-chemical and other process industries).

Number of revolutions or on/off operations.

There are no absolutes in terms of right or wrong choice for a particular product, as any sensible choices will be closely correlated. However, the use of the variable that most closely corresponds to the failure mechanism at work will, in any particular data set, minimize the uncertainty in parameter estimation, as described later. In practise it is not always immediately apparent what the most appropriate choice of parameter is, and in these cases the correct choice will only be found by trial and error.

The value of t at which failure occurs is, of course, unknown in advance – it is a random variable that necessitates a probabilistic rather than deterministic approach. If failures were predictable, as in the case of the oft-quoted "One Hoss Shay" of Oliver Wendel Holmes (that lasted for "a year and a day") then the subject would become trivial. The key to the modelling of the 'lifetimes' of a series of components of the same type is the concept of the 'lifetime probability distribution', as shown in Fig. 6.1 (see also section 1.2(a) and Fig. 1.3).

Without, for the moment, ascribing any particular shape to this distribution, $f(t)$ is defined (as in section 1.2(a)) as follows

$f(t)$ = Probability Density Function
and the total area under the curve = 1

This area is referred to as the 'distribution function' denoted by $F(t)$, that is

Distribution function, $F(t)$
= Area under the curve (Fig. 6.1)

At any value of t, such as t^*, say, the probability that the component has failed at or before this time is the area under the curve to the left (shown shaded in Fig. 6.1), that is

Probability that component has failed $F(t^*)$
= Area shaded under curve (Fig. 6.1)

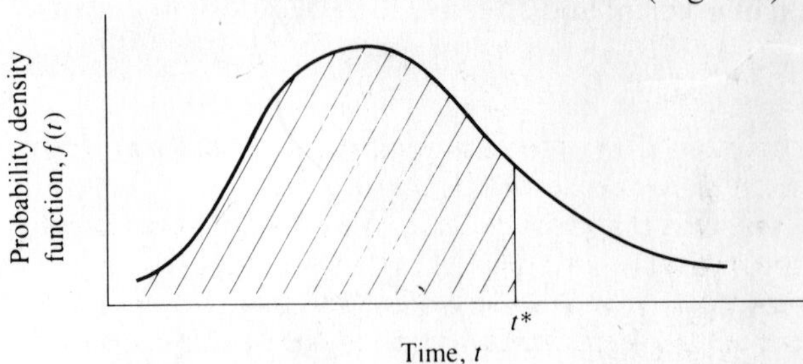

Fig. 6.1. Lifetime probability distribution

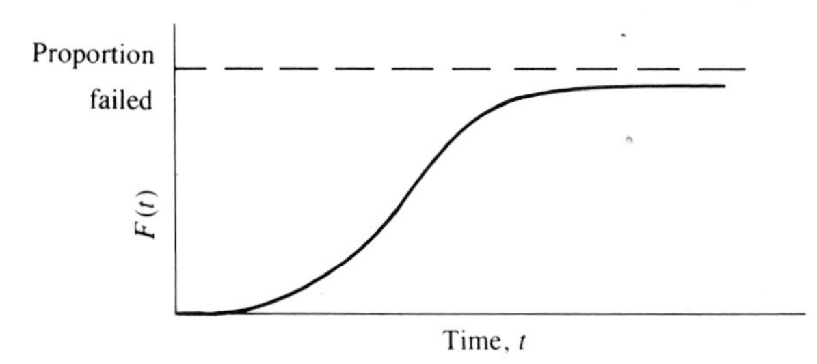

Fig. 6.2. Distribution function

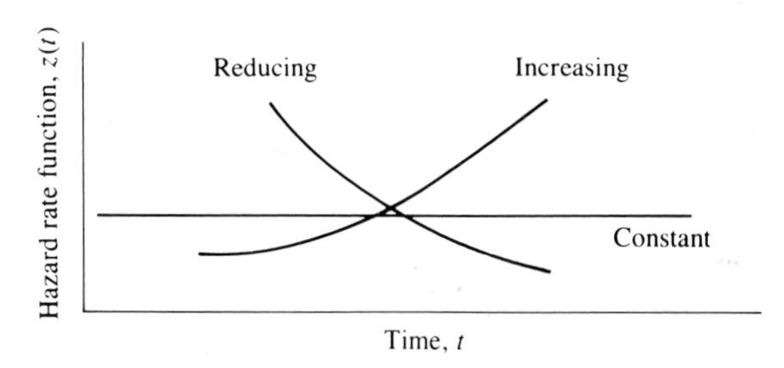

Fig. 6.3. Typical behaviour curves for the hazard function

It should be noted that

$$\text{at } t = 0 \qquad F(0) = 0$$

$$\text{and at } t = \infty \qquad F(\infty) = 1$$

$$\text{and, more generally,} \quad F(t) = \int_0^t f(u)\, du \qquad (6.1)$$

The graph of the Distribution Function, $F(t)$, against t depends on the shape of the probability distribution (Probability Density Function), but will be of the general form shown in Fig. 6.2.

We will make particular use of this function when considering probability plotting and related subjects, as in section 6.4 below.

There are two other related functions that are useful in describing component reliability.

(a) Reliability Function, R(t)

This is the probability that a component has survived to time t, and is simply the complement of the distribution function; that is

$$R(t) = 1 - F(t)$$

Note that, in Fig. 6.1

$R(t)$ = Area under the curve to the right of t*

(b) Hazard Rate Function, z(t)

This is a very useful intuitive measure of component behaviour, and is defined as a *rate* (that is, probability per unit time).

At time t, the Hazard Rate Function, $z(t)$, is the probability of failure, *given* survival to time t.

In terms of Fig. 6.1, it is the ordinate value of $f(t)$ at t, divided by the area to the right of t, for any value of t. That is

$$z(t) = \frac{f(t)}{R(t)} \qquad (6.2)$$

Typical behaviour curves of the Hazard Rate Function, $z(t)$, are shown in Fig. 6.3.

For some components, the Hazard Rate Function may assume a more-or-less constant value, that is, the likelihood of a failure is independent of the age of the component. This is often true in the case of electronic components and other components where failures are due to random causes unrelated to component age. Such failures are themselves 'random' in the strict statistical sense; beware, however, of terminology in use that uses 'random' to denote failures that are rare or unexplained, but not necessarily of constant hazard. Constant hazard is widely assumed when it is not appropriate as it has the attraction of being mathematically much simpler than the alternatives.

(1) *Increasing hazard* – the component gets more likely to fail as it gets older. This will occur in any situation where use of the product degrades it, for example, corrosion, wear, fatigue, and so on. As this applies to many engineering components it suggests that the assumption of constant hazard is, in many circumstances, at least questionable.

(2) *Reducing hazard* – the component gets less likely to fail as it gets older. A common manifestation of this is the component that is initially highly stressed due to misalignment and the stress is reduced as the component 'beds in'.

(c) Cumulative hazard, H(t)

A further related conceptual function that does not have any obvious intuitive practical meaning but which will also be found useful in plotting methods for data analysis is that of Cumulative Hazard, denoted $H(t)$. This is simply the area under the Hazard Rate Function curve, as shown in Fig. 6.4. That is

Cumulative Hazard Function, $H(t)$
= Area under Hazard Rate Function, $z(t)$, curve

As the Cumulative Hazard Function is an integral of

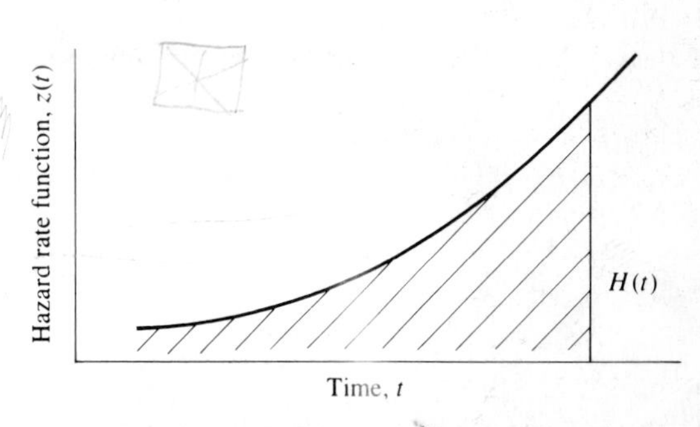

Fig. 6.4. Relationship between cumulative hazard and Hazard Rate Functions

the Hazard Rate Function, $z(t)$, over the time period to t, it is no longer a rate and it can be shown that it is, in fact, simply related to the Distribution Function, $F(t)$, by

$$H(t) = \log_e \left\{ \frac{1}{1 - f(t)} \right\}$$

from which

$$F(t) = 1 - e^{-H(t)} \tag{6.3}$$

6.3 The Weibull distribution

So far, no particular shape has been assigned to the lifetime distribution. Whilst data analysis without so doing is possible, it is usual to assign some specific function to $f(t)$, thereby constraining it to a particular shape or family of shapes. Some well known functions include the exponential distribution (which describes the constant hazard case) and the normal and log-normal (which are restricted to increasing hazard). The most widely used model in component reliability analysis, however, is the Weibull distribution (**5**). This originated in fatigue studies, and it is of particular practical significance as it was derived empirically. It has several features which make it attractive to practising reliability engineers and which account for its very wide use, viz:

- flexibility – it can deal with increasing, constant, and reducing hazard;
- mathematical simplicity;
- amenability to graphical analysis;
- a demonstrated ability to fit most lifetime data better than most of its potential competitors.

In its simple 'two parameter' form the Weibull model is most simply described by its Distribution Function

$$F(t) = 1 - e^{(t/\eta)^{\beta}} \tag{6.4}$$

Where

η is the scale parameter, known as the 'characteristic life'. It is the value of t at which there is an approximately 2/3 probability that the component will have failed (strictly, the probability is $1 - 1/e = 0.632$).

β is a shape parameter which is related to the behaviour of the hazard function.

For $\beta = 1$, the Hazard Rate Function is constant (that is, equivalent to the exponential distribution).

For $\beta > 1$, the Hazard Rate Function is increasing, and the bigger it is the more rapidly is it increasing. It is unusual to encounter β values greater than 4.

For $\beta < 1$, the Hazard Rate Function is reducing. The smaller it is, the more rapid is the reduction. β must be greater than zero, and it is unusual to encounter values lower than 0.5.

The constants η and β are parameters that specify which

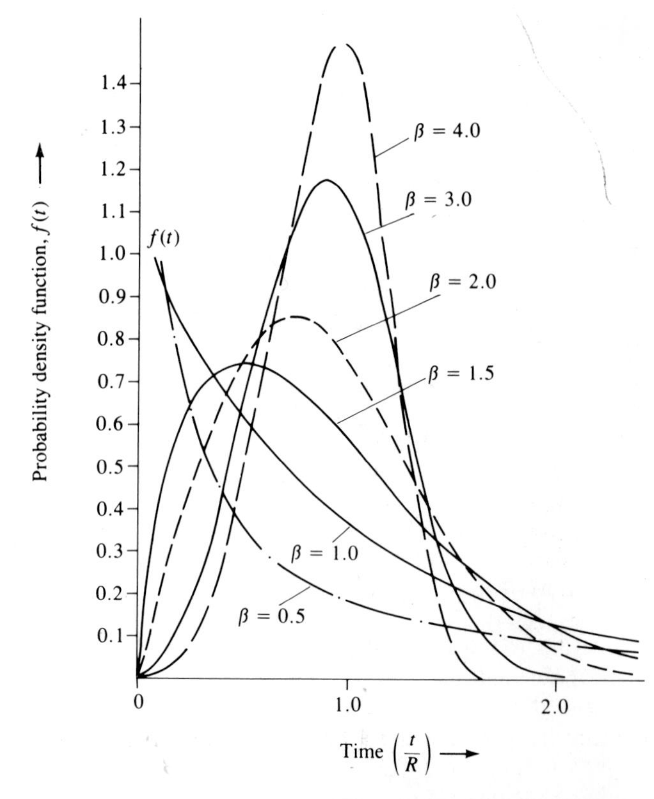

Fig. 6.5. The Weibull Probability Density Function, $f(t)$, for different values of the shape parameter, β

particular member of the 'family' of functions or curves will describe the data. Component reliability analysis based on the Weibull model reduces to a process of estimation of these parameters.

The behaviour of the Weibull Probability Density Function, $f(t)$, for a selection of values of β is shown in Fig. 6.5, and the behaviour of the Weibull Hazard Rate Function, $z(t)$, for a selection of β values is shown in Fig. 6.6.

To summarize.

(1) The Shape Parameter, β, defines the form of the Hazard Rate Function, $z(t)$, and indicates whether it is increasing, constant, or decreasing.

(2) The Characteristic Life, η, is an indirect measure of overall reliability – the larger its values the longer the component is expected to survive.

Note also that procedures are available for obtaining estimates of the mean and standard deviation of the Weibull distribution from estimates of β and η. These values may be required for planned replacement and stock control models.

6.4 Weibull analysis

This is the process of obtaining estimates for β and η from observed failure data. There are both graphical and analytical methods available, but analysis here will concentrate on the former, which, in addition to being

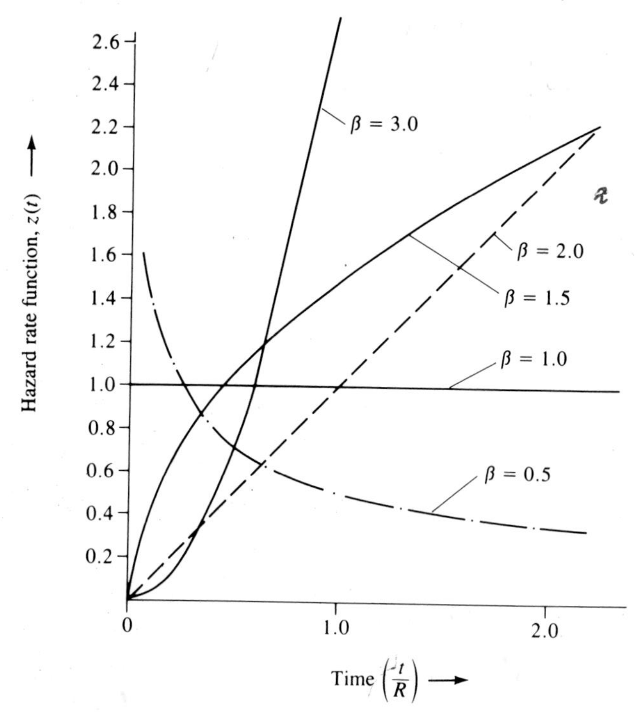

Fig. 6.6. The Weibull Hazard Rate Function, *z(t)*, for different values of the shape parameter, *β*

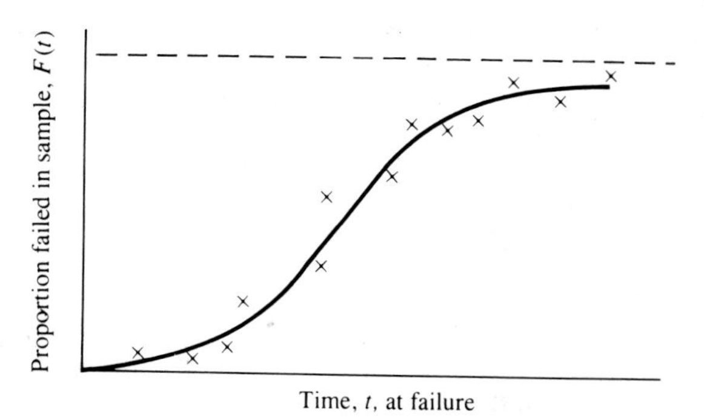

Fig. 6.7. Distribution function for failures in a given sample

particularly applicable to the Weibull model, has the attraction of producing additional subjective information about the data beyond that included in the estimates alone.

Graphical analysis itself needs considering under two headings – one referring to what is known as 'complete' data (where all the items in the sample have failed), and the other to 'censored' data which includes unfailed components. Censored data is defined more fully later.

In either case the principle is the same – the proportion failed in the observed sample at a particular t value is used as an estimate of the value of the Distribution Function, $F(t)$, at that time. If these estimates are plotted, at each observed failure time, a graph similar to Fig. 6.7 results.

The objective is to find the parameters of the Weibull Distribution Function that most closely fit the plotted points. To circumvent the difficulty of fitting curves, and also of determining which of the infinite number of possible curves are in fact Weibull distribution functions, it is usual to use graph paper with axes transformed in such a way that the plot will be linear.

There are several such papers commercially available – the most widely used in the UK is a version due to Nelson (**6**) and produced by Chartwell, as shown in Fig. 6.8.

Data is analysed as if it consisted of a life test on a number (n) of similar components that were tested until k of them had failed (where $k \leq n$). Even if, as is usual in practice, that is not the way in which the results were obtained, the procedure operates as if this were the case, and is as follows.

(a) Write down the k failure ages of the components in increasing order of magnitude, viz.

t_1 = smallest ordered age at failure
t_2 = second smallest ordered age at failure
t_i = ith ordered age at failure
t_k = largest ordered age at failure

(b) At each t_i, calculate the corresponding estimate of the distribution function $\hat{F}(t_i)$

This is usually given by

$$\hat{F}(t_i) = \frac{i - 0.3}{n + 0.4} \qquad (6.5)$$

where

i = failure number
n = total number of components in the sample

At first it might seem more sensible to use the simpler expression i/n for $\hat{F}(t_i)$. This would, in fact, give an overestimate because our observation is not at a random time, but at a failure time where $\hat{F}(T_i)$ makes a step function increase. The equation given is known as Benards equation and is derived in reference (**7**). It is one of several widely used expressions, and experience shows it to give satisfactory results. The comparison of this expression with other approaches is considered in detail by Barnett (**8**).

This procedure is applicable to data that is either complete (that is, all items in the sample have failed ($n = k$)) or 'singly censored' data where all ($n - k$) unfailed items have survived for times equal to or greater than t_k. For the case of multiply censored data where these conditions are not met a slightly more complex procedure is required – see section 6.5.

(c) Plot the data on Weibull graph paper as shown in Fig. 6.8 and draw the best straight line through the data points.

(d) Construct a perpendicular to the plotted line which goes through the estimation point ringed at the top left hand corner of the Weibull graph paper. Where this perpendicular cuts the β estimation line gives the estimate for β.

(e) The scale parameter estimator is a horizontal broken line at an ordinate value (cumulative failures) of approx. 63 per cent. Where this broken

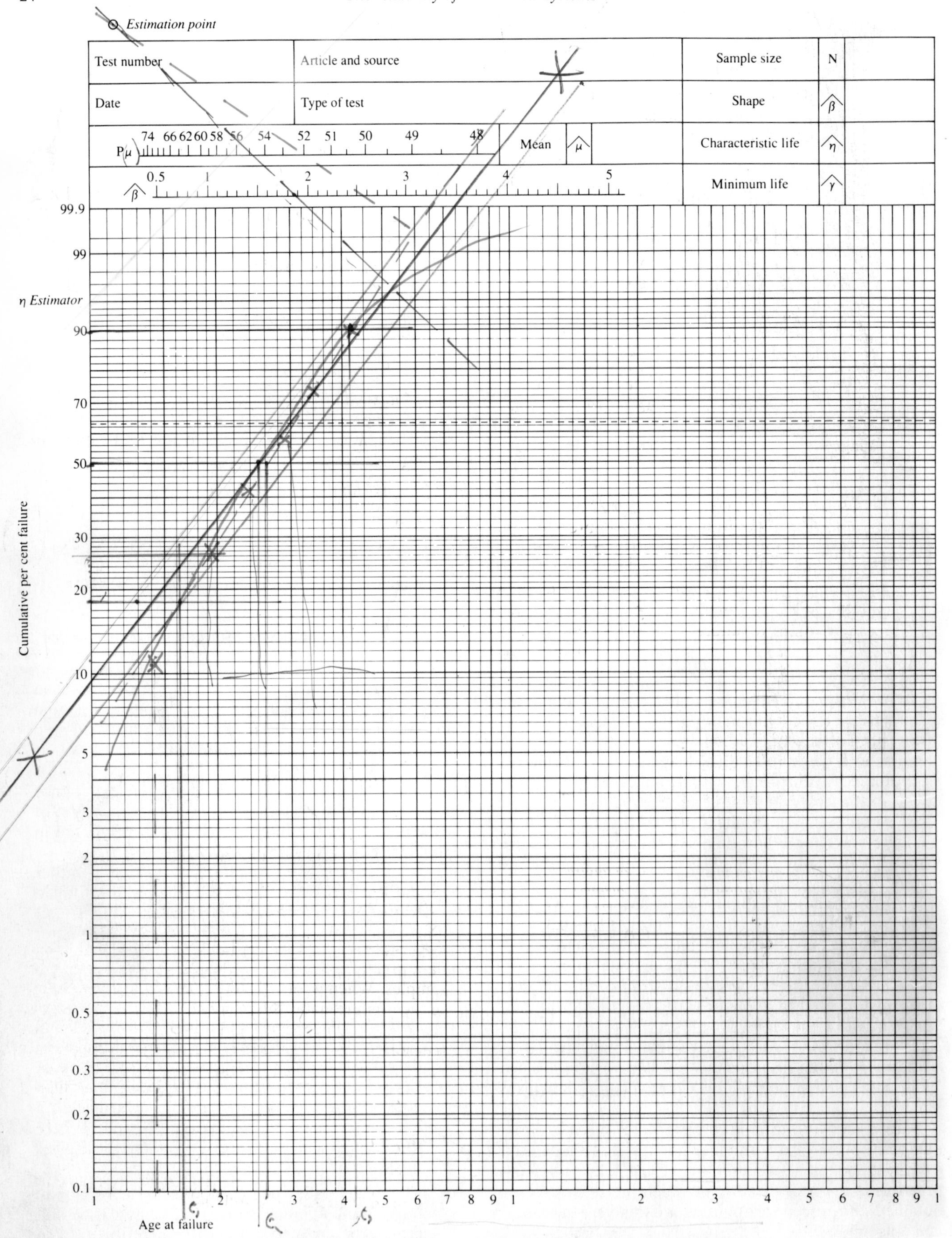

Fig. 6.8. Weibull distribution function graph paper, as produced by Chartwell

line cuts the plotted line drop a vertical line to the horizontal time axis, and this gives an estimate of the characteristic life $\hat{\eta}$.

(f) The estimate of the percentage of components failed at the mean life $P\mu$ is given where the perpendicular from the plotted line cuts the $P\mu$ line.

(g) The percentage of components failed at the mean life $P\mu$ is then plotted on the vertical ordinate of the Weibull paper and a horizontal line drawn to cut the plotted line. Where the cut takes place a vertical line is dropped to the horizontal time ordinate to give the estimated mean life of the sample of components.

The procedure is illustrated by the following example.

Example 6.1

The data in Table 6.1 refers to a life test on a sample of 20 switches. The quoted life is the number of operations at which failure occurred. The unfailed switches were removed after 4500 operations. Obtain the Shape Parameter, β, the Characteristic Life, η, the Mean Life, and the Standard Deviation.

Note that the sample size, n, is 20 and not 15, the number of failures. This data is 'singly censored' in that the test was stopped at an arbitrary time before all the switches had failed. There is no particular difficulty in handling this data because all the censoring times (that is, time at which unfailed items are removed) are greater than the largest time to failure. When this is not the case we have 'multiply censored' data that is much more difficult to deal with.

(a) The data is plotted on Weibull graph paper and is shown in Fig. 6.9, and a perpendicular constructed through the Estimation Point.

Table 6.1. Life test data on sample of 20 switches

Failure No.	*No. of operations to failure*	*Estimate of probability of failure at time* t
(i)	(t)	$\hat{F}(t) = \dfrac{i - 0.3}{n + 0.4}$
1	430	0.034
2	900	0.083
3	1090	0.132
4	1220	0.181
5	1500	0.230
6	1910	0.280
7	1915	0.328
8	2250	0.377
9	2600	0.426
10	2610	0.475
11	3000	0.525
12	3390	0.574
13	3430	0.637
14	3700	0.672
15	4050	0.721

(b) Shape Parameter

From Fig. 6.9 where the perpendicular cuts the β scale

$$\beta = 1.62$$

and from this estimate it is clear that the Hazard Rate Function, $z(t)$, is increasing.

(c) Characteristic Life

From the Fig. 6.9 the η estimator cuts the plotted line and a vertical dropped from this cut to the horizontal line gives

$$\hat{\eta} = 3600 \text{ operations}$$

(d) Mean Life

From Fig. 6.9 the perpendicular crosses the percentage failed at the mean life, P_μ, scale to give

$$P_\mu = 57 \text{ per cent}$$

Plotting this value on the vertical axis, draw a horizontal from this to cut the data plot, and drop a vertical line from the cut to the horizontal axis, giving

$$\text{Estimated Mean Life, } \hat{\mu} = 3100 \text{ operations}$$

(e) Standard Deviation

An approximate value of the Standard Deviation is given by multiplying the Characteristic Life by a factor, B, from Table 6.2 interpolating where necessary.

Table 6.2. Multiplication factor for standard deviation

β	1.0	1.5	2.0	2.5	3.0	3.5	4.0	5.0
B	1.0	0.61	0.51	0.37	0.32	0.28	0.25	0.21

Therefore

$$\text{Standard deviation} = 3600 \times 0.59$$
$$= 2124 \text{ operations}$$

6.5 Multiple censored data and hazard plots

Section 6.4 has dealt with the analysis of either complete data (where all the items in the sample have failed) or 'singly censored' data (where all the survival ages are equal to or greater than the largest time to failure). Such data usually arises from life tests on specially procured samples. When we are dealing with the analysis of field service failure data, the problems of multiple random censoring often appears. In this case, unfailed lifetimes (censorings) are mixed with the failure times, which makes estimation of $\hat{F}(t_i)$ less obvious than previously.

Consider, as a simple illustration of multiple censored data, the following example.

Example 6.2

A sample of five motor vehicles were tested for a particular mode of failure dependent on vehicle usage. Data from the test was as follows.

Vehicle No. 1: Currently at 22 000 km without failure.

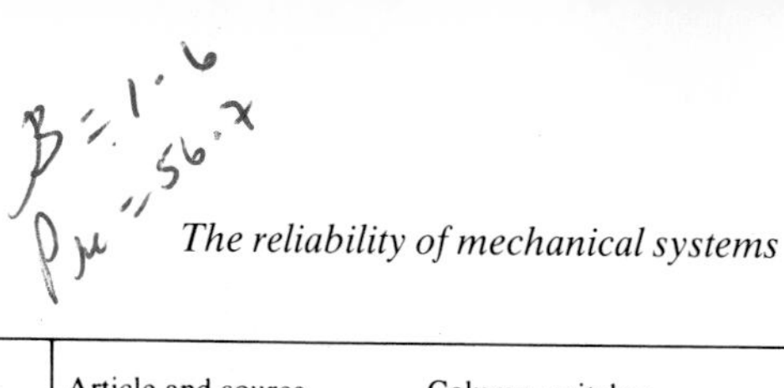

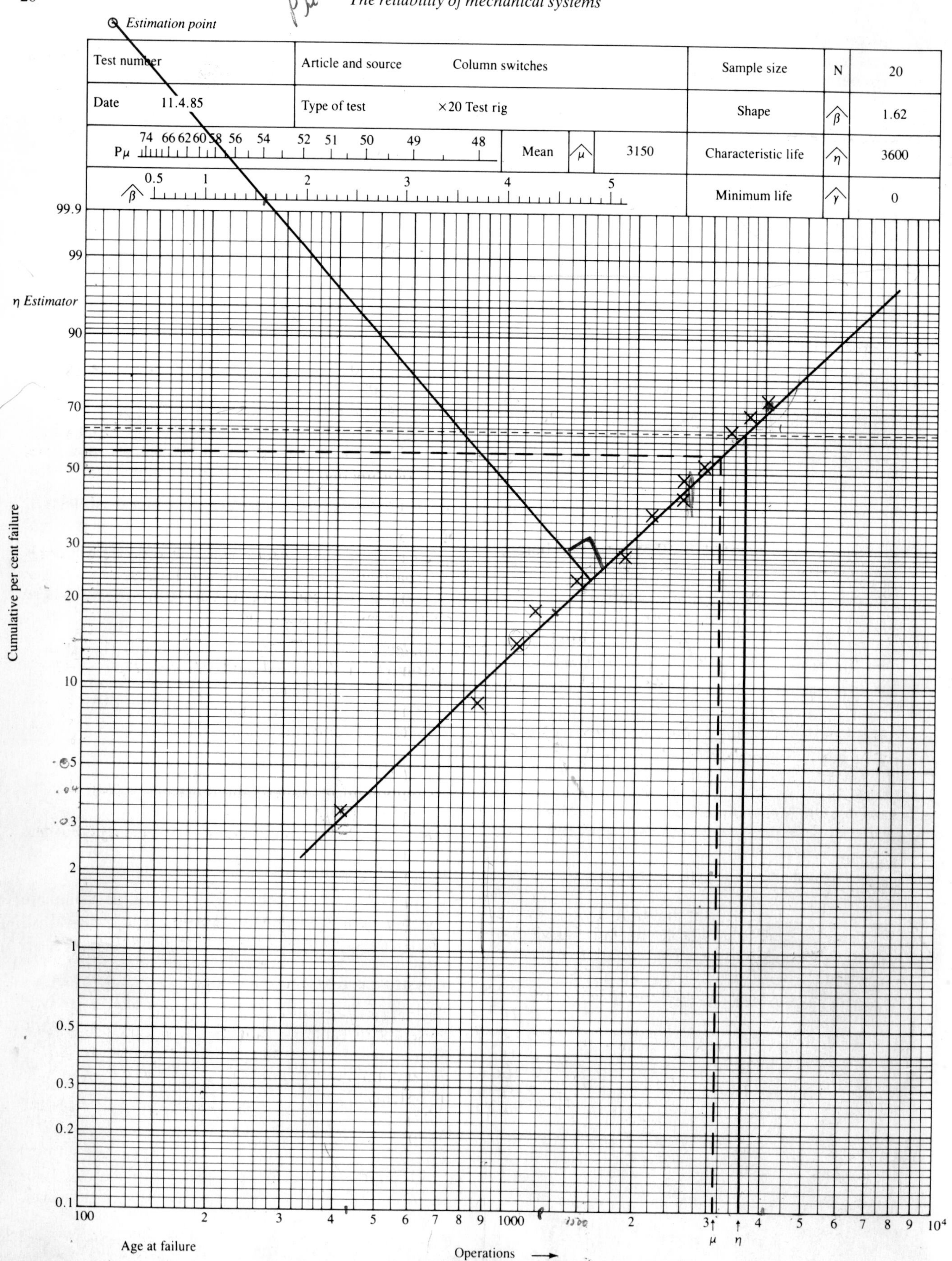

Fig. 6.9. Weibull plot of data for Example 6.1

Vehicle No. 2: Failed at 40 000 km. Removed from test.
Vehicle No. 3: Failed at 5100 km. Removed from test.
Vehicle No. 4: Destroyed in accident at 9500 km.
Vehicle No. 5: Failed at 15 000 km. Removed from test.

Putting these events in increasing order of distance travelled

5100	Failure	(F1)
9500	Censoring	(C1)
15000	Failure	(F2)
22000	Censoring	(C2)
40000	Failure	(F3)

Sample size $(n) = 5$. Failures $(k) = 3$

Calculate the mean order of these events

(a) For the first item there is no problem – it is the first failure in a sample of 5.

(b) For the failure at 15000 km there are two possibilities:

(i) it is the second failure (that is, C1 fails after F2)

or (ii) it is the third failure (that is, C1 fails before F2)

By considering all the arrangements of F2, it is possible to arrive at a 'mean order' of 2.25. Similarly, F3 could be the third, fourth, or fifth failure, depending on what would have eventually happened to C1 and C2 – its mean order is in fact 4.125.

Calculation of mean order can be tedious with large samples. Other methods have been suggested (for example, Kaplan–Meir estimates), but the 'Hazard Analysis' due to Nelson (**9**) will be used here. This is a method that has achieved widespread use in reliability analysis and is particularly applicable to the Weibull distribution.

At any time at which a failure occurs, the sample estimate of the Hazard Rate Function is simply the number of failures occurring at that time divided by the number in the sample available to fail (survivors) immediately before that time. The cumulative sum of these values will give a sample estimate of the Cumulative Hazard Function, $H(t)$, as described in section 6.2, Fig. 6.4.

Table 6.3. Results from Example 6.3

Time T_i	*No. of Failures* (x)	*Survivors* (s)	*Hazard Rate Function* $\hat{z}(t)$ (= x/s)	*Cumulative Hazard Function* $\hat{H}(t)$ $(= \Sigma\{ln\ (t)\})$
5100	1	5	0.200	0.200
9500	—	4	—	—
15000	1	3	0.333	0.533
22000	—	2	—	—
40000	1	1	1.000	1.533

Example 6.3

Using the data from Example 6.2 calculate the hazard rate function, $z(t)$, and the cumulative hazard function, $H(t)$.

For Weibull analysis there are now two possibilities.

(1) Plot $\log_e \hat{H}(t)$ against t – this will under the Weibull model give a straight line of slope β.

or

(2) Convert the $\hat{H}(t)$ estimates into $\hat{F}(t_i)$ using

$$\hat{F}(t_i) = 1 - e^{-H(t_i)}$$

that is

t_i	$\hat{H}(t_i)$	$\hat{F}(t_i)$
5100	0.200	0.181
15000	0.533	0.413
40000	1.533	0.784

These $\hat{F}(t_i)$ values can now be plotted on Weibull plotting paper in the usual way.

Two further examples illustrate these points more fully.

Example 6.4

Three repairable systems were monitored for failures of a particular component of which there are two per system (denoted component A and component B). The resulting raw data is as follows:

System 1
Component A – failed (and replaced) at 3780 and 6362 hrs
Component B – failed (and replaced) at 4885 hrs
System currently at 8000 hrs

System 2
Component A – failed (and replaced) at 1040 hrs
Component B – no failures
Both components replaced as precautionary measures at 5000 hrs
System currently at 6500 hrs

System 3
Component A – failed (and replaced) at 2180 hrs
Component B – failed (and replaced) at 2777 and 5082 hrs
System currently at 6950 hours

The data is re-organised in terms of component lives and presented in Table 6.4. Note that component censorings occur at the current system lives (as they are still operating and unfailed) and also at the planned replacements in System 2.

These results are plotted in Fig. 6.10 from which the Weibull parameters are estimated as:

Shape Parameter, $\beta = 2.0$
Scale Parameters, $\eta = 4200$
Mean Life, $\mu = 3750$

From these estimated parameters the following conclusions about the reliability of the component can be drawn.

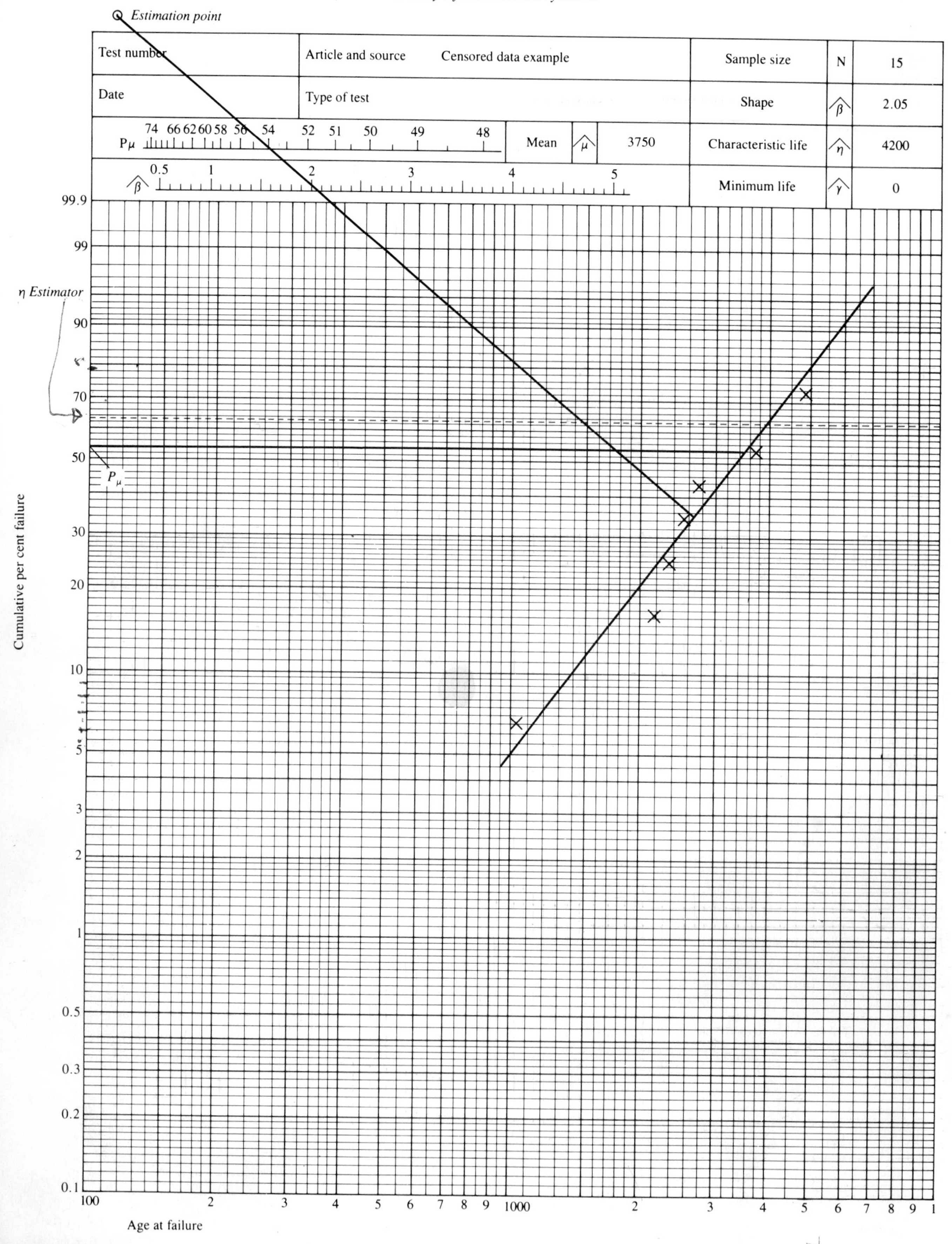

Fig. 6.10. Weibull plot of data for Example 6.4 (Table 6.4)

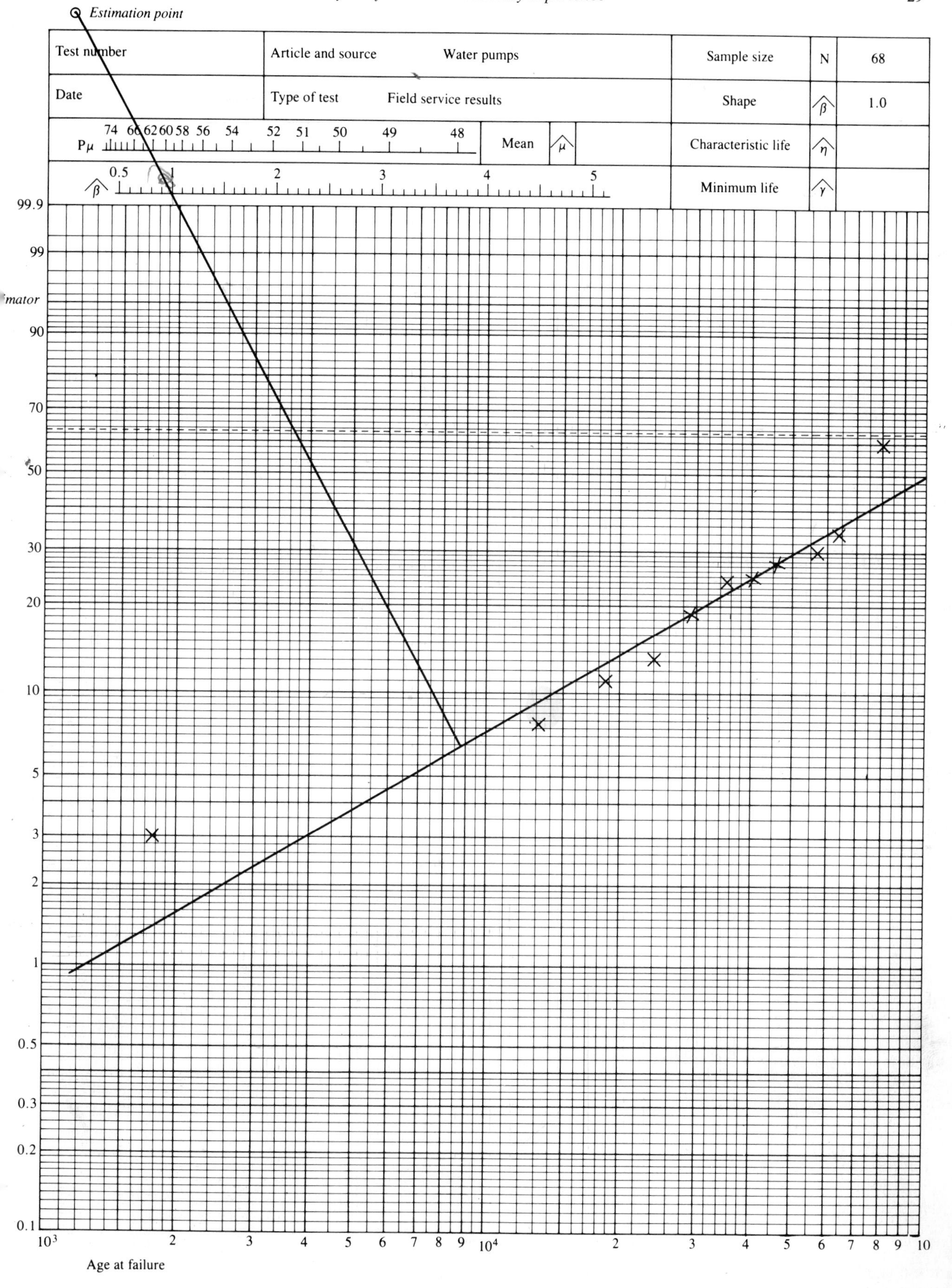

Fig. 6.11. Weibull plot of data for Example 6.6 (Table 6.5)

Table 6.4. Multiple Censored Data for Example 6.4

Component	*Age* (t_i) *of each component at failure*	*Failure or censoring*	*Survivors*	*Hazard rate function* $\hat{z}(t_i)$	*Cumulative hazard function* $\hat{H}(t_i)$	*Corresponding value of* $\hat{F}(t)$
2A1	1040	F	15	0.0667	0.0667	0.0645
2B2	1500	C	14	—	—	—
2A3	1500	C	13	—	—	—
1A3	1638	C	12	—	—	—
3B3	1868	C	11	—	—	—
3A1	2180	F	10	0.1000	0.1667	0.1535
3B2	2305	F	9	0.1111	0.2778	0.2426
1A2	2582	F	8	0.1250	0.4028	0.3316
3B1	2777	F	7	0.1428	0.5456	0.4205
1B2	3115	C	6	—	—	—
1A1	3780	F	5	0.2000	0.7456	0.5256
2A2	3960	C	4	—	—	—
3A2	4770	C	3	—	—	—
1B1	4885	F	2	0.5000	1.2456	0.7122
2B1	5000	C	1	—	—	—

(1) As a component fails and is replaced a new component with a new start to its life (age) is created. The exercise involved a total of 15 components, 5 for each system.
(2) The components are denoted by system (1, 2, or 3), by component (A or B) and by which component in the series after failure or replacement (1, 2, or 3). Thus the first component in System 1 in the A series is denoted 1A1 with an age of 3780 hours. The third component in System 1 after two component failures is denoted 1A3 with a component age of 1638 hours (that is, 8000 − 6362). The second component in System 2 in the B series, after replacement of the first component before failure, is denoted 2B2 with a component age of 1500 hours (6500 − 5000), and so on.
(3) Age (t_i) is the age in hours of each individual component at failure, replacement or at survival to the current life of the system.

(1) The component exhibits moderate wear-out characteristics.
(2) The expected life of any component is 3750 hours.
(3) 63 per cent of the components are expected to have failed at 4200 hours.

Example 6.5

The second example shows lifetimes of a component in the engines of a series of diesel trucks. It includes failures and planned replacements of the component and current unfailed ages (on trucks that are currently operating). These have been re-arranged as in the previous example to give ordered events (in terms of distance covered (km) at failure). The only difference from the previous example is that the data is grouped into class intervals. The simplifying assumption is made that failures and censorings are spread evenly throughout each interval – from this assumption we plot the km values as the mid-point of each interval, and determine the hazard from the average of the survivors at the beginning and the end of each interval (that is, the hazard in the class). The data is shown in Table 6.5 and a Weibull plot is given in Fig. 6.11.

The results in this case (from Fig. 6.11) show that

$$\hat{\beta} = 1$$

which indicates a constant Hazard Rate Function, $z(t)$. The intersects for the Characteristic Life, η, and the Mean Life, μ, are off scale and the horizontal axis needs to be extended to obtain the necessary readings.

Finally it is very important to include all the censorings (survivors) that occur in the analysis. It is incorrect to ignore them as this will always result in an under-estimation of the Characteristic Life, η.

It should be noted that computer software is available to carry out the graphical analysis, and typical computer output for a Weibull analysis is shown in Fig. 6.12.

6.6 Non-graphical analysis – maximum likelihood estimates

It is possible to avoid the plotting procedures described in the sections by substituting algebraic methods that will produce the estimates required. This is easily done for complete data (where, for example, the mean and variance of the data provide estimates of the corresponding population values) but is more difficult when there is any censoring. Cohen (**10**) has produced equations for maximum likelihood estimates of the Shape Parameter, β, and the Characteristic Life, η, for which a simple computer program can be written.

Whilst the use of such a program will give 'instant' values of the estimates, it is recommended that plots are made in addition, as they provide much useful information, particularly when complications arise (such as those described in section 6.7 below).

Table 6.5. Truck component data for Example 6.5

	1		*2*	*3*	*4*	*5*
(Thousands)	*Failures* (x)	*Censorings*	*Survivors* (s)	*Hazard* (z(t))	*Cumulative hazard* (H(t))	$\hat{F}(t)$
0–5	0	1	68	0	0	0
5–10	2	4	67	0.0313	0.0313	0.031
10–15	3	3	61	0.0517	0.0830	0.080
15–20	2	3	55	0.0381	0.1211	0.114
20–25	1	1	50	0.0204	0.1415	0.132
25–30	3	1	48	0.0652	0.2067	0.187
30–35	3	0	44	0.0706	0.2773	0.242
35–40	1	3	41	0.0256	0.3029	0.261
40–45	1	7	37	0.0303	0.3332	0.283
45–50	0	2	29	0	—	—
50–55	1	4	27	0.0408	0.3740	0.312
55–60	1	7	22	0.0555	0.4296	0.349
60–65	0	6	14	0	—	—
65–70	0	3	8	0	—	—
70–75	2	1	5	0.5714	1.0010	0.632
75–80	0	1	2	0	—	—
80–85	0	1	1	0	—	—
>85	0	0				
	20	48				

(1) For interval 25–30

Failures, $x = 3$

All survivors, $s = \frac{48 + 44}{2} = 46$

Hazard rate function, $z(t) = \frac{3}{46} = 0.0652$

(2) For interval 70–75

Failures, $x = 2$

All survivors, $s = \frac{5 + 2}{2} = 3.5$

Hazard rate function, $z(t) = \frac{2}{3.5} = 0.5714$

(3) The Weibull probability plot in this case will be of $F(t)$ against the mid-point of each time interval, that is, 2500, 7500, 12500, and so on. The results are plotted in Fig. 6.11.

6.7 Some common problems in Weibull analysis

There is a temptation to be over critical of departures from a straight line in a Weibull plot. It must, however, be remembered that we are not dealing with a deterministic physical relationship (such as a 'Hookes law' plot), but a statistical distribution function in which errors are unavoidable. Indeed, too-good a fit indicates that the data could be suspected of having been 'adjusted'.

There are, however, examples of clear departure from linear plots that sometimes occur in practice. Two commonly occurring ones are as follows.

(a) *Location parameter (monotonic curved plot)*

The appearance of a monotonic curve indicates that the origin of the t_i values is not zero, but some non-zero constant known as the 'Location Parameter'. Conventional symbols are γ or t_0.

A positive value of γ describes a lifetime distribution as shown in Fig. 6.13(a) and will result in a curved Weibull plot as shown in Fig. 6.13(b).

This situation arises when there is some interval (from $t = 0$ to $t = \gamma$) during which failures cannot occur. An example might be the erosion of a protective coating.

A negative value of γ describes a lifetime distribution as shown in Fig. 6.14(a) and will result in a curved Weibull plot as in Fig. 6.14(b). This could be caused, for example, by a 'shelf life' problem where failures can occur before the start of normal service life; for example, pitting corrosion of bearings while held as spares before being taken into service.

In either case, the existence of a Location Parameter should not be accepted without some sensible engineer-

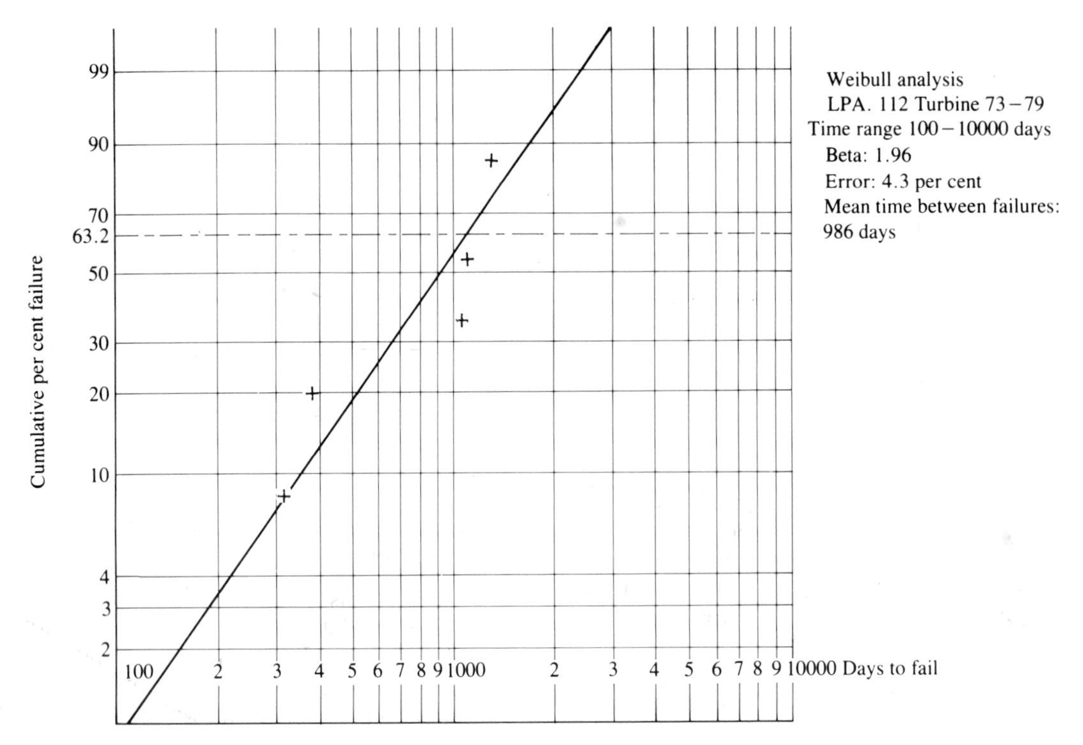

Fig. 6.12. Typical computer output for a Weibull analysis

ing explanation. It should also be viewed with suspicion if it is linked with a low value of the shape parameter, β (less than about 1.5), which would imply a sudden discontinuity in the distribution. If there is a plausible justification, the procedure for estimating the location parameter, γ, is as follows.

(1) Draw three equispaced parallel horizontal lines on the plot, as shown in Fig. 6.15 (the 'equispacing' is in terms of actual distance on the paper, not the $F(t)$ scale). The distance between the top and bottom lines is not critical, but they should just encompass all the plotted points.

(2) Drop verticals from the intersection of these lines with the plotted curve, at t_1, t_2, and t_3, as shown.

(3) The Location Parameter, γ, is estimated by

$$\gamma = t_2 + \frac{(t_3 - t_2)(t_2 - t_1)}{(t_3 - t_2) - (t_2 - t_1)} \tag{6.6}$$

When this estimate has been obtained, it should be subtracted from all the observed failure and censoring times, and the data re-plotted on Weibull graph paper. The re-plot will give a straight line from which the Shape Parameter, β, can be estimated in the usual way.

(b) Competing failure modes ('bent' plot)

A Weibull plot such as Fig. 6.16 indicates that two independent, competing causes of failure are present, with different Weibull parameters.

What must *not* be done in this case is simply to estimate the values for the Shape Parameter, β, and Characteristic Life, η, for each part of the line as if the other did not exist – this will give completely incorrect values. The correct procedure is to produce separate plots for each mode of failure, treating the other one(s) as censorings.

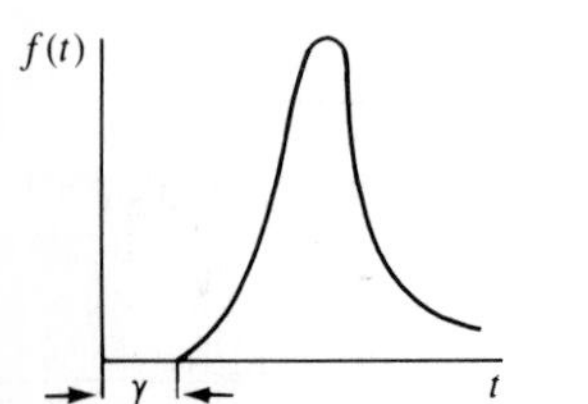

(a) Probability distribution (b) Weibull plot

Fig. 6.13. Positive location parameter

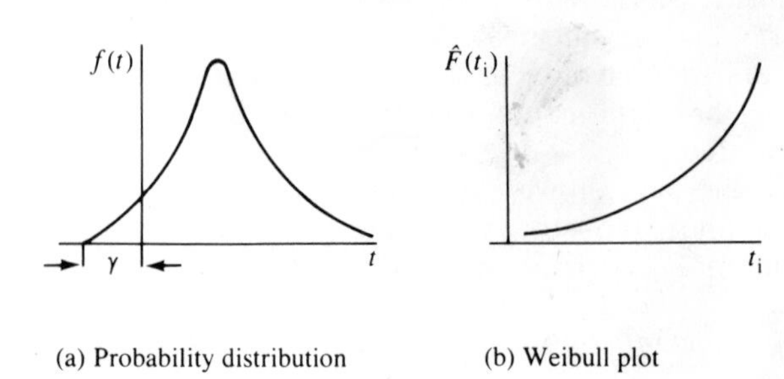

(a) Probability distribution (b) Weibull plot

Fig. 6.14. Negative location parameter

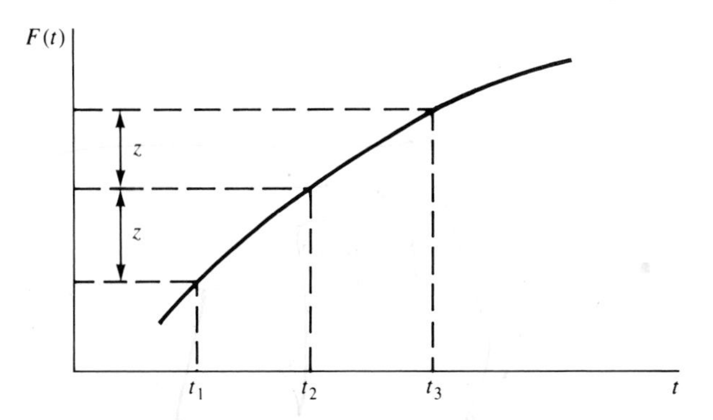

Fig. 6.15. Graphical estimation of γ

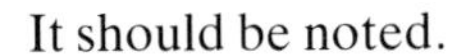
It should be noted.

Even if the dichotomy between the lines is clear, it is difficult to separate the data points referring to each line from the plot alone – and dangerous to try. Identification of the failure modes can only be satisfactorily achieved by an engineering investigation of the failed components.

The plot cannot be relied upon to demonstrate two modes even if they are present.

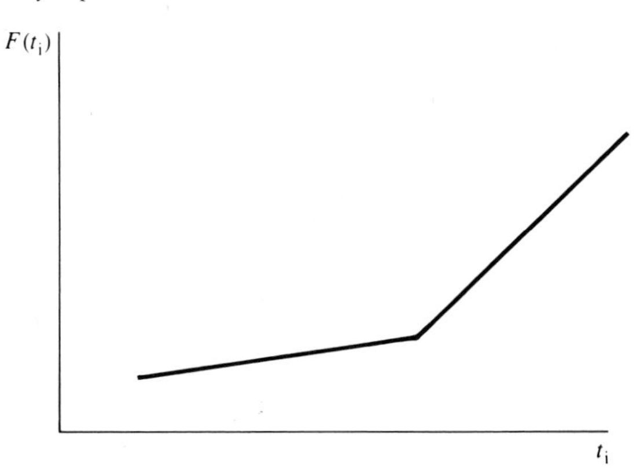

Fig. 6.16. Two competing failure modes

The same principles apply to more than two modes, but the mixing obscures them even further.

Mixing of distribution will always increase the variance of the data, which will reduce the apparent value of the Shape Parameter, β. Important 'wear-out' modes are often not detected because the data contains other failure modes which obscure them.

CHAPTER 7. ANALYSIS OF IN-SERVICE EXPERIENCE FOR REPAIRABLE SYSTEMS

7.1 Introduction

This section deals with repairable systems where the occurrence of a failure does not necessarily cause the 'end of life' of the system; rather, the failed element is repaired or replaced to restore the system to its operating state.

On this basis, for such a system we can observe the system elapsed times at which failures occur.

Defining

y_i = system age at the ith failure

t_i = interval between $(i-1)$th and ith failure

The behaviour of the system can be represented as in Fig. 7.1.

We will make some initial simplifying assumptions.

(1) The repairs that take place at each y_i are 'perfect', that is, they restore the failed equipment to its 'as good as new' condition.

(2) The failures occur at random, but at a constant underlying failure rate. This implies that the failures are the result of a Homogeneous Poisson Process (HPP) giving inter-failure times, t, that are exponentially distributed (and the number of failures in specified time intervals have a Poisson distribution.

Under the Homogeneous Poisson Process (HPP) model, over any 'packet' of elapsed time totalling T, the expected number of failures is λT. The probability of observing some arbitrary number of failures, x, is given by the Poisson distribution

$$\text{Probability of } x \text{ failures, } P(x) = \frac{e^{-m}m^x}{x} \qquad (7.1)$$

where

$m = \lambda T$

Note. Circumstances when these assumptions are justified, and the procedures to adopt when they are not, are discussed in section 7.4

For a system that complies with these assumptions, a plot of cumulative failures, $N(t)$, against system age would produce a result as shown in Fig. 7.2, the expected relationship being a straight line. The slope of this line, $(d/dt)N(t)$ is called the 'system failure rate' (usual symbol λ), and its reciprocal is called the 'Mean Time Between Failures' (MTBF) (usual symbol θ).

7.2 A note on 'Failure Rate'

'Failure Rate', as described above relates specifically to the failures per unit time of a repairable system – it must be distinguished from the concept of hazard rate functions for non-repairable components that form elements within such a system.

Such components may exhibit:

reducing hazard – in which case they become less likely to fail as they get older;

constant hazard – in which case their probability of failure is unaffected by their age;

increasing hazard – in which case they become more likely to fail as they get older.

They may even move from one category to another as they age – for a discussion of the mechanisms resulting in particular hazard patterns see, for example, Carter (**11**).

The essential point in considering repairable systems is that, providing the assumption of 'as good as new' applies, for a complex system of failure rate will tend to a constant value as system age y increases, irrespective of the Hazard Rate Functions of the components in the system. The occurrence of increasing or reducing system failure rates (as discussed in 7.4) is a completely separate issue from that of increasing or reducing Hazard Rate Functions (as discussed in Chapters 2 and 6.

7.3 Reliability analysis under constant failure rate

A useful effect of the assumption of constant failure rate is that it is possible to simply aggregate system life, t, over any number of systems irrespective of system age to give a total elapsed time, T.

If, in this elapsed time T, we have observed x failures, then

$$\text{Estimated failure rate} \quad \hat{\lambda} = \frac{x}{T} \qquad (7.2)$$

$$\text{Estimated MTBF} \quad \hat{\theta} = \frac{T}{x} \qquad (7.3)$$

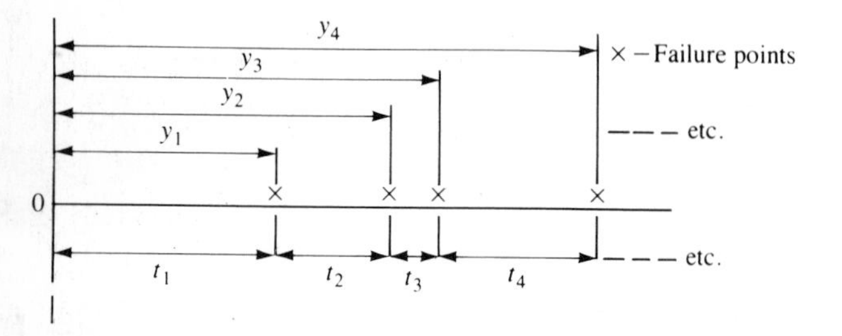

Fig. 7.1. A repairable system

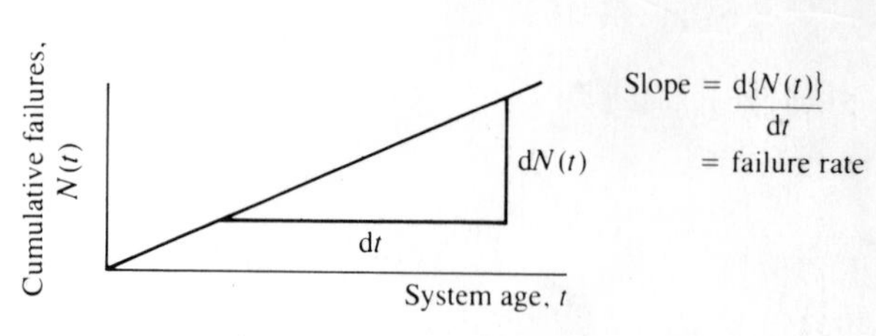

Fig. 7.2. System failure rate

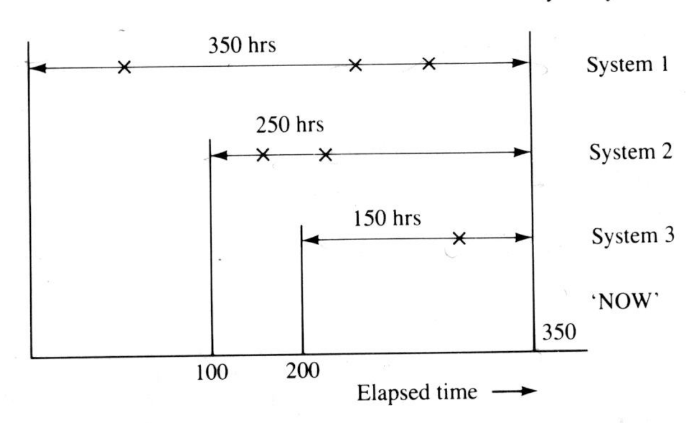

Fig. 7.3. Repairable system example

Example 7.1

Figure 7.3 represents the history of three repairable systems. System 1 starts its life at time 0. System 2 starts its life when System 1 has been operating for 100 hours. System 3 starts its life when System 1 has been operating for 200 hours (and System 2 for 100 hours). Failures are denoted by x and there have been six failures to date.

Currently Systems 1, 2 and 3 have, respectively, been operating for 350, 250 and 150 hours.

The current total elapsed time, T, is

$$T = 350 + 250 + 150$$
$$= 750 \text{ hours}$$

Number of failures, $x = 6$.

Therefore, for all three systems, the Estimated Failure Rate, $\hat{\lambda}$, is given by

$$\hat{\lambda} = \frac{x}{T} = \frac{6}{750}$$
$$= 0.008 \text{ failures/hour}$$

and the estimated MTBF, $\hat{\theta}$, is given by

$$\theta = \frac{T}{x} = \frac{750}{6}$$
$$= 125 \text{ hours}$$

Example 7.2

Suppose we know that a particular type of system has an MTBF of 100 hours. What is the probability of getting one failure in a total of 100 hours operation?

$$\text{Failure Rate, } \lambda = \frac{1}{\theta} = \frac{1}{100} = 0.01;$$

$$T = 100 \text{ and } \lambda T = 1.0$$

From equation 7.1

$$\text{Probability of failures } P(1) \frac{e^{-m}\, m^x}{x} = \frac{e^{-1} x\, 1^1}{1}$$
$$= e^{-1}$$
$$= 0.37 \text{ (or 37 per cent)}$$

That is, there is a 37 per cent chance of 1 failure in 100 hours.

(This is, incidentally, a result that should be of particular interest to the breed of engineers who view the MTBF as if it were a guaranteed failure-free life – in fact, there is only about one in three chance of surviving to the MTBF without failure!)

Probabilities of 1, 2, 3, etc., failures can be calculated from the same expression for this (or any other) value of system elapsed time, T.

(a) Confidence intervals for failure rate

In reliability estimation, it is usual to put a confidence limit on the estimate which usually answers the question; 'Suppose that, because of natural variation in sample results, there is an optimistic test result; how bad might the true reliability really be?'

This is most easily explained in terms of the expected number of failures in our test (m) if the Failure Rate were λ. That is $m = \lambda T$. If a high value of λ ($= \lambda_U$) is defined such that, for this value, the probability of getting the observed number of failures, x, or lower, in the observed total time, T, is some small value, α, then λ_U is known as the upper $(1 - \alpha)$ confidence limit for the failure rate, λ.

That is, if it is stated that 'the true value of λ does not exceed λ_U', there is a probability $(1 - \alpha)$ that the statement is correct.

Although it is usually of less interest to consider 'how good might the reliability be?', a lower $(1 - \alpha)$ confidence limit for λ ($= \lambda_L$) is similarly defined such that the probability of obtaining x or more failures is α.

If it is required to consider MTBFs rather than failure rate, simply note that $\theta_L = 1/\lambda_U$ and $\theta_U = 1/\lambda_L$.

Expressions for confidence limits for failure rate and MTBF are tabulated in Table 7.1.

Table 7.1. Confidence limits for MTBF and failure rate

	(1 − α) Confidence limit	*Time terminated*	*Failure terminated*
Failure rate	Upper (λ_u)	$\frac{\chi^2_{\alpha,2(x+1)}}{2T}$	$\frac{\chi^2_{\alpha,2x}}{2T}$
	Lower (λ_L)	$\frac{\chi^2_{(1-\alpha),2x}}{2T}$	$\frac{\chi^2_{(1-\alpha),2x}}{2T}$
MTBF	Upper (θ_u)	$\frac{2T}{\chi^2_{(1-\alpha),2x}}$	$\frac{2T}{\chi^2_{(1-\alpha),2x}}$
	Lower (θ_L)	$\frac{2T}{\chi^2_{\alpha,2(x+1)}}$	$\frac{2T}{\chi^2_{\alpha,2x}}$

(1) The symbol χ^2 denotes a chi-square variate, where the first suffix represents the right hand tail area of the chi-square distribution, and the second suffix the mean value of the distribution (known as the 'degrees of freedom'). For example, $\chi^2_{0.1,15}$ would be the χ^2 value shown in Fig. 7.4(a) (which is numerically equal to 22.307), and $\chi^2_{0.95,20}$ would be the value in Fig. 7.4(b) (numerically equal to 10.581.

(2) The time terminated case refers to the situation where the test is stopped at some pre-determined time. The failure terminated case refers to the case when the test is terminated at the occurrence of a failure.

Example 7.3

Bloggs Aerospace tested a sample of 15 amplifiers supplied by Dryjoint Electronics Limited. All fifteen were put on test and, after 500 hours testing, seven amplifiers had failed at 48, 90, 132, 240, 335, 380, and 412 hours. Determine the lower 90 per cent confidence limit for the MTBF.

Total test time for failed units is given by

$$T_F = 48 + 90 + 132 + 240 + 335 + 380 + 412$$
$$= 1637 \text{ hours}$$

Total test time for unfailed units is given by

$$T_U = 8 \times 500$$
$$= 4000 \text{ hours}$$

Therefore, total test time for all units is

$$T = T_F + T_U = 1637 + 4000$$
$$= 5637 \text{ hours}$$

Therefore, from equation (7.3) the estimated MTBF is

$$\text{Estimated MTBF, } \hat{\theta} = \frac{5637}{7}$$
$$= 805 \text{ hours}$$

From Table 7.1 the lower 90 per cent confidence limit for MTBF (time terminated) is

$$\theta_L = \frac{2T}{\chi^2_{\alpha,2(x+1)}}$$
$$= \frac{2 \times 5637}{\chi^2_{0.1,2(1+7)}} = \frac{11274}{\chi^2_{0.1,16}}$$

From the Table for χ^2 in reference (**12**)

$$\chi^2_{0.1,16} = 23542$$

Therefore

$$\theta_L = \frac{11274}{23542}$$
$$= 479 \text{ hours}$$

That is, the MTBF is estimated to be 805 hours, and the lower 90 per cent confidence limit is 479 hours. (In other words, if we state that the true MTBF is not lower than 479 hours, there is a 90 per cent chance of the statement being correct.)

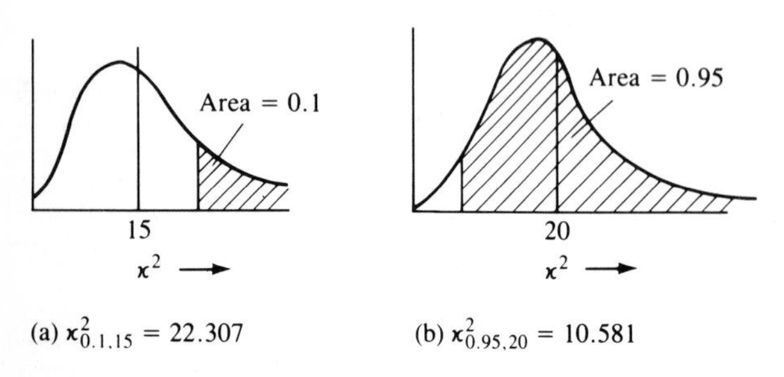

Fig. 7.4. Examples of the Chi Square distribution

Example 7.4

Bloggs now continue testing until all the amplifiers from Example 7.3 have failed, the failures occurring at 503, 660, 662, 1037, 1380, 1413, 2078, and 2212 hours. What is your revised estimate of the confidence limit for MTBF?

Total test time for failed units from Example 7.3 is

$$T_1 = 1637 \text{ hours}$$

Total test time for failed units from continuation test is

$$T_2 = 503 + 660 + 662 + 1037 + 1380 + 1413$$
$$+ 2078 + 2212$$
$$= 9945 \text{ hours}$$

Total test time

$$T = T_1 + T_2$$
$$= 1637 + 9945$$
$$= 11\,582 \text{ hours}$$

Therefore, estimated MTBF

$$\theta = \frac{11582}{15}$$
$$= 772 \text{ hours}$$

From Table 7.1 the lower 90 per cent confidence limit for MTTF (failure terminated) is

$$\theta_i = \frac{2T}{\chi^2_{\alpha,2x}}$$
$$= \frac{2T}{\chi^2_{0.1,30}}$$

From the table for χ^2 in reference (**12**)

$$\chi^2_{0.1,30} = 40256$$

Therefore

$$\theta_L = \frac{23164}{40256}$$
$$= 575 \text{ hours}$$

That is, the estimated MTBF is 772 hours, and there is a 90 per cent chance that it will not be less than 575 hours.

(b) A cautionary note on zero failures

It should be noted that, whilst the occurrence of zero failure in any amount of testing gives an estimated failure rate of zero (that is, an infinite MTBF), it is still possible to calculate a lower confidence limit for failure rate.

Example 7.5

Suppose a test is carried out for 100 hours and there are zero failures. The resulting estimate of zero for the failure rate should be tempered with the fact that the upper 95 per cent confidence limit, from Table 7.1, is

$$\lambda_U = \frac{\chi^2_{\alpha,2(x+1)}}{2T} = \frac{\chi^2_{0.05,2}}{200}$$

From the table for χ^2 in reference (**12**)

$$\chi^2_{0.05,2} = 5991$$

Therefore

$$\lambda_U = \frac{5991}{200}$$
$$= 0.03 \text{ failures/hour}$$

or that the lower 95 per cent confidence limit for MTBF, θ_L, is

$$\theta_L = \frac{1}{\lambda_L}$$
$$= 33.4 \text{ hours}$$

In other words, although the estimated MTBF for a zero failure rate is infinity, there is a 95 per cent chance that it will not be less than 33.4 hours, or conversely a 5 per cent chance that it will be less than or equal to 33.4 hours.

This could quite likely be viewed as unsatisfactory. It also leads into the more general question, 'How much testing is needed to show that the reliability is adequate?', which is considered in section 7.3(c).

(c) Reliability demonstration

Reliability (usually in the guise of MTBF) is often quoted as a specification requirement for a system, together with the confidence level with which it is to be demonstrated that the requirement is met.

This is achieved by producing a demonstration test result whose probability of occurrence if the MTBF is equal to or lower than the required value is (1 − confidence level).

The test is defined by T, the accumulated test time, obtained from a re-arrangement of the appropriate expression (for lower MTBF limit, timed terminated in Table 7.1). That is

$$T = K\theta_L$$

where

$$K = \frac{\chi^2_{\alpha,2(x+1)}}{2}$$

θ_L = lower confidence limit for MTBF

K is a factor by which the required MTBF is multiplied to give the required test time, knowing the required confidence level $(1 - \alpha)$ and the number of failures that have been observed – shown, for convenient reference, in Table 7.2 for some conventional values of confidence level (expressed as percentages).

Example 7.6

The following clause is included in the specification for a compressor, 'It shall be demonstrated with a 90 per cent confidence that the MTBF is not less than 150 hours'.

Here a test result is needed such that, if the MTBF is 150 hours, the probability of obtaining the number of failures that do in fact occur on test, or fewer, is not more than 10 per cent.

For a satisfactory test time

$$T = K\theta_L$$

Therefore, from Table 7.2, for zero failures to meet the specification requirement the test would need to accumulate T_0 hours without failure, where

$$T_0 = 2.3 \times 150$$
$$= 345 \text{ hours}$$

If a failure occurred before this time the test would have to be extended to T_1

$$T_1 = 3.89 \times 150$$
$$= 583 \text{ hours} \quad \text{and so on}$$

(d) Acceptance tests

The problem with demonstration as described above in section 7.3(c) is that, whilst a reliable system will be accepted, an unreliable one will not be rejected and the accumulated hours will simply fall progressively further behind the target as testing continues. For instance, in Example 7.6 there might be:

one failure after 210 hours, in which case it would be necessary to continue to test up to 583 hours;

a second failure at 330 hours, in which case it would be necessary to continue to test up to 798 hours (that is, 5.32 × 150);

a third failure at 425 hours in which case it would be

Table 7.2. MTBF multiplication factors for reliability demonstration
(Derived from the χ^2 table in reference (**12**))

Confidence level (%)	*Number of failures*					
	0	*1*	*2*	*3*	*4*	*5*
50	0.70	1.68	2.67	3.67	4.67	5.67
75	1.39	2.69	3.92	5.06	6.27	7.42
80	1.61	2.99	4.28	5.52	6.72	7.91
90	2.30	3.89	5.32	6.68	8.00	9.27
95	3.00	4.74	6.30	7.75	9.15	10.60
99	4.60	6.64	8.41	10.04	11.60	13.11

necessary to continue to test up to 1002 hours (that is, 6.68×150);

and ad infinitum.

Clearly, on this evidence, this product is so unreliable that it will never meet the reliability requirements in its present state of development. However, human nature being what it is, there is a temptation to keep on testing in the hope that the test requirement will be met. To avoid this, it is useful to introduce a procedure whereby a 'reject' decision can be reached (meaning 'take the system off test, go away, and improve it').

The way to define a suitable acceptance test is to specify two levels of failure rate, as follows.

(1) A low value of the failure rate, λ_1, at which we have a probability of $(1 - \alpha)$ of accepting the product.

(2) A higher value of the failure rate, λ_2, at which we have a probability of β of accepting the product. (Here it is convenient to work in terms of failure rate rather than MTBF.)

α and β are small values (typically 0.05, but we can make them any value less than 0.5) known, respectively, as the producers and consumers risks. The Producers Risk (α) is the risk of wrongly rejecting a product which has a 'satisfactory' Failure Rate, λ_1. The Consumers Risk (β) is the risk of wrongly accepting a product which has the 'unsatisfactory' Failure Rate, λ_2. The ratio λ_2/λ_1 is known as the 'discrimination ratio'.

Various testing procedures have been proposed, but a very efficient and widely used one is the 'Sequential Probability Ratio Test' (SPRT). In this type of test, a log is kept of the accumulated test time (T) and the total number of failures ($N(T)$), and they are plotted as in Fig. 7.2 (except that the test is not necessarily restricted to a single system, so we use T instead of t). The accept/reject criteria are given by drawing on the plot a pair of parallel lines, as shown in Fig. 7.5, of equations

For the Accept line, $T = h_0 + s(N(T))$

For the Reject line, $T = -h_1 + s(N(T))$

To avoid the risk of testing for a long time without crossing either line (which might happen if λ was about half way between λ_1 nd the λ_2) the test envelope is truncated, as shown in Fig. 7.5.

Expression for parameters for sequential probability ratio test.

$$h_0 = \frac{\ln\left(\frac{1-\alpha}{\beta}\right)}{\lambda_2 - \lambda_1}$$

$$h_1 = \frac{\ln\left(\frac{1-\beta}{\alpha}\right)}{\lambda_2 - \lambda_1}$$

$$S = \frac{\ln(\lambda_2/\lambda_1)}{\lambda_2 - \lambda_1}$$

where α, β, λ_1, and λ_2 are as defined above.

To calculate the truncation criterion, r (which is in fact the rejection number for the fixed single sample test plan of the form 'test for time T, reject for r or more failures'), find the lowest integer that satisfies the inequality

$$\frac{\lambda_2}{\lambda_1} \neq \frac{\chi^2_{\beta,2r}}{\chi^2_{(1-\alpha),2r}}$$

The accept line is truncated at $T = Sr$, and the reject line is truncated at $T = r$.

As an alternative, use can be made of published test plans based on these principles – the most usual source of such is the US military specification MIL-STD-781D.

Example 7.7

Design a test that has a 95 per cent chance of giving an 'accept' decision if the MTBF is 150 hours, but only a 10 per cent chance of accepting for an MTBF of 50 hours.

Substitution in the expressions above gives the results

$$h_0 = 169; \quad h_1 = 217; \quad s = 82.4; \quad r = 8$$

and maximum test time $= s \times r = 82.4 \times 8 = 659$ hours.

The graph of this test plan is shown in Fig. 7.6.

7.4 The assumption of constant failure rate

The foregoing material has assumed throughout that the failure rate exhibited by the system (λ) is constant, that is, that the plot of cumulative failures against total elapsed time is a straight line, as in Fig. 7.2.

This assumption will, in general, be valid, if all the elements in the system exhibit a constant hazard rate

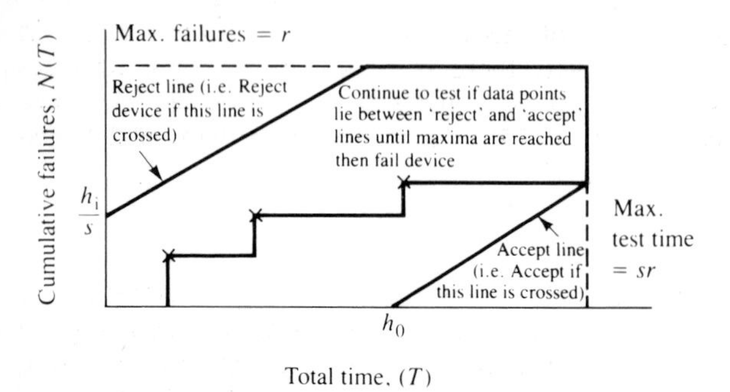

Fig. 7.5. Sequential probability ratio test

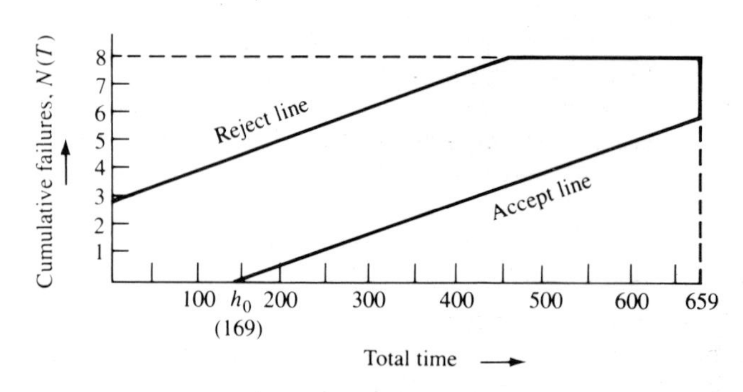

Fig. 7.6. Example of SPRT

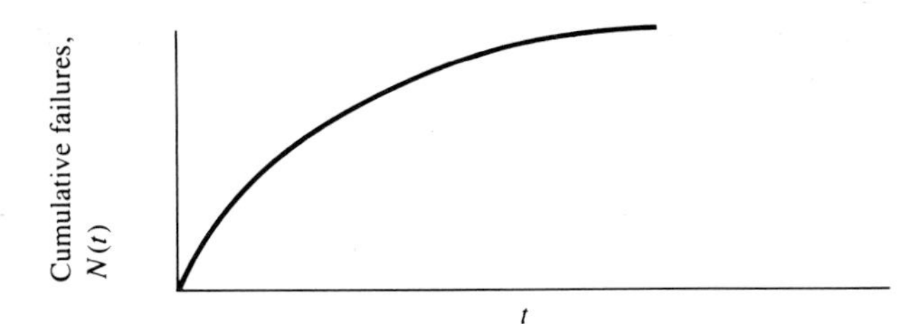

Fig. 7.7. Graph of cumulative failures showing reliability growth

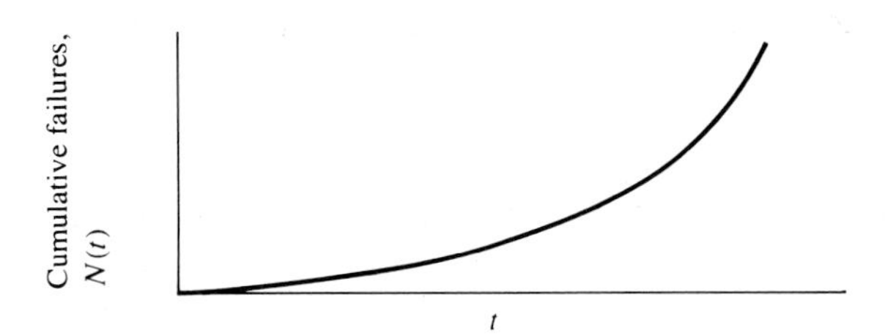

Fig. 7.8. Graph of cumulative failures showing reliability deterioration

function. This is generally true in the case of electronic components, but in the case of mechanical components this is not the case – non-constant Hazard Rate Functions are commonplace, usually increasing, but also including the 'initial rise followed by a fall' typically found in fatigue failures. If there is a large number of components, with no dominant failure mode, there will be a 'mixing' effect of the failures and replacements so that the overall failure rate (which can be approximated by the sum of the individual component hazards) will soon settle down to an essentially constant value, as in Fig. 7.2. This is sometimes referred to as a 'pseudo-constant failure rate' to distinguish it from the situation arising from constant hazard components.

This constant failure rate pattern now depends on the further assumption of 'as good as new' replacement of failed components; that is, the replacement item will have the same lifetime distribution as the item it replaced (which is not the same as saying it will fail at the same age). There are some circumstances in which we expect this *not* to be true, viz.

(1) The consequences of engineering development make the replacement item superior to the failed item. In this case we would expect it to last longer. If this happens to several of the failure modes, we would expect the times between successive failures to tend to increase, giving a result as in Fig. 7.7.

This particular form of non-constant failure rate has been the subject of extensive further analysis under the guise of 'reliability growth modelling' – see section 7.6.

(2) Conversely, the replacement item may be 'worse than new'. This usually results from the replacement item having received a less-than-perfect repair. As an example, a fatigue crack in a casting may be repaired by welding, but other potential cracks are unaffected and their propagation is related to the start of life on the casting, not the repair time.

Reliability deterioration of this sort will give a result as in Fig. 7.8.

Reliability growth or deterioration must not be confused with reducing or increasing hazard – they are independent of each other, and any combination is possible, as discussed in section 7.2.

Example 7.8

The following example is used to illustrate several techniques in the rest of this section. It refers to the results of prolonged trial during the development testing of four prototype motor vehicles.

Vehicle No.	*Cumulative vehicle km at which failures occurred (thousands)*
1	2, 17, 19, 26, 38, 57, 101, 141, (150)
2	0.3, 10, 12, 15, 41, 87, (120)
3	3, 5, 11, 14, 23, 51, 66, 79, 113, (120)
4	0.9, 5, 6, 9, 17, 19, 53, 71, (100)

(Figures in parentheses are km at the end of trial for each vehicle)

There are 31 failures at *total* elapsed distances of: 1.2, 3.6, 8, 12, 20, 20, 24, 36, 40, 44, 48, 56, 60, 68, 68, 76, 76, 92, 104, 152, 164, 204, 212, 228, 264, 284, 316, 348, 403, 439, and 481, and the total distance covered is 490 (all in units of 1000 km.

Note that this assumes all vehicles have a common time origin – effectively, that they were in the same state of development at time 0 – and that they travelled at the same speed. On this basis, the first failure (in vehicle 2) occurred at a distance of $t = 0.3$, so the total distance for all four vehicles was $0.3 \times 4 = 1.2$. Other values were calculated similarly, making due allowance for the progressive 'dropping-out' of the trial of vehicle 4, 3, and 2.

A plot of cumulative failures against distance is shown in Fig. 7.9. The fact that this data exhibits a reducing failure rate is obvious from the plot, and no further confirmation of this fact is necessary. In less clear-cut cases, however, there may be a need for more confirmatory analysis – suggested techniques are the Laplace trend test (**13**) or a reliability growth model, discussed later.

7.5 Reliability growth analysis

For a repairable system, the widespread assumption of constant failure rate depends on the assumption of 'good as new' replacements, that is, the replacement for a failed element is of the same standard (with the same expected lifetime) as the failed item. The actual achieved

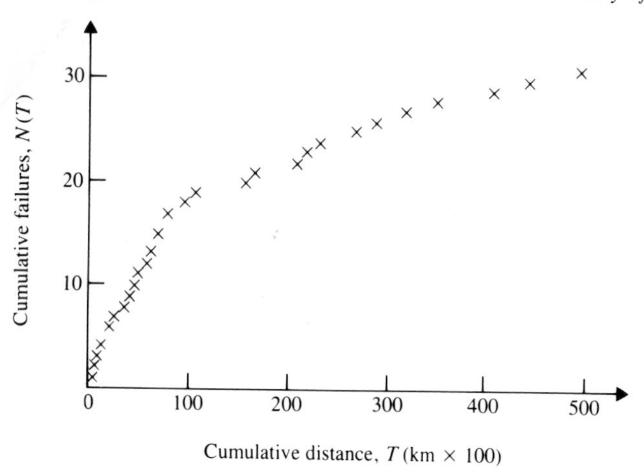

Fig. 7.9 Cumulative failure plot

lifetime is, of course, subject to sampling variation, and may be greater or smaller than that of its predecessor – it is its expected (or mean) value that is the same.

If this were to be true, it would in fact be a criticism of the engineering development process. Every failure has an engineering cause, and in an ideal world the product design should be improved on the basis of information obtained from failures until their causes are eliminated. This ideal world is rarely, if ever, achieved in practice because the law of diminishing returns would make it uneconomic, but what one can hope for is to eliminate many of the more persistent causes of premature failure, leaving a 'base' level of unavoidable wear-out and fatigue failures, plus, perhaps, a few 'nuisance' failures of low level and minor consequence.

As such engineering development takes place, failed items will tend to be replaced not by 'As Good As New' items, but by 'better than new', and the failure rate curve will show a reducing slope, as in Fig. 7.7 in section 7.4.

The test given in Example 7.7 can be used to see if such growth is taking place, with a general rule that the larger the (negative) test statistic, the more rapid the growth.

7.6 Reliability growth models

Some more specific models have been proposed which assume specific forms for the shape of the curve in Fig. 7.6, and the simpler models in most widespread use are based on an exponential 'learning curve'. The first, and best known, of these is due to Duane (**2**). It was derived empirically from observation of the reliability improvement patterns of a number of electronic and mechanical systems. From this has been developed the AMSAA model which is a different way of expressing the same basic logic model as Duane, and other more sophisticated models (such as the SIL and IBM models).

The object of all these models is to quantify the rate of reduction of failure rate, and to predict the further development effort needed to meet specific targets.

PART THREE

A basic approach to reliability assessment for mechanical process systems

CHAPTER 8. USES OF RELIABILITY ASSESSMENT

8.1 Introduction

In Parts One and Two the basic principles, concepts, and philosophy of reliability engineering have been introduced, and the basic techniques which are widely used for the analysis of reliability experience for components (but which can also be used for equipment and systems) have been demonstrated. In parts Three and Four some of the techniques of reliability assessment are explained. Part Three is concerned with the basic techniques of reliability assessment, which, although comparatively simple, may often be usefully applied. Chapter 9 introduces block and logic diagrams and shows how they can be applied to the analysis of simple series and parallel systems. Chapter 10 extends this analysis to process systems which include built in redundancy or on-line standby equipment or components, so that not all equipment or components have to operate continuously to ensure that the system operates satisfactorily. A great many practical problems can be investigated to a sufficient degree of accuracy using these methods, and a grasp of their principles and limitations will enable an engineer to establish which problems require a more complex analysis.

8.2 The need for reliability assessment

In almost all specifications for mechanical plant systems there will be either an explicit or implied statement to the effect that: 'The plant shall be designed to operate safely and reliably'. In some instances this may mean little more than that the plant should function with as little inconvenience and cost to the manufacturer or supplier as possible until the end of the guarantee period; in others it may mean that the plant must operate continuously between annual planned shutdowns; and in a few instances that it is imperative that the plant does not fail when it is required to operate in an emergency. The level of safety and reliability required will often be reflected in the length and complexity of the specification itself. However, while these general statements of reliability requirements play their part in putting pressure on suppliers to improve reliability, once specific values for safety and reliability can be quantified from plant specifications, then reliability assessment becomes necessary in order that the supplier can demonstrate, at the design stage, that the plant is likely to meet the specification requirements.

The need for reliability assessment arises from three primary concerns of plant users, namely, economics, safety, and project viability.

(a) Economics

Plant users and operators have long recognised that the capital costs of purchase and installation are not the only costs which effect a plant's economic viability, and that in choosing between competing tenders an assessment must also be made of likely operating and maintenance costs. However, at the time when tenders are submitted, while capital costs may be fixed, operating and maintenance costs are based on predictions of likely plant availability and behaviour. It is essential, if meaningful comparisons between competing tenders are to be made, that these predictions are based on rational assessments of plant and equipment reliability. The ways in which operating and maintenance costs are related to reliability have been described in Chapter 1. Here it is only necessary to highlight that, in economic assessments, the most important parameter to be derived is usually plant availability. This has already been defined in Chapter 3 as the proportion of time for which a component, an equipment, or a system is capable of performing its duty, whether it is actually performing its duty or is in some standby role.

Plant designers and manufacturers have become aware that reliability assessment can help to demonstrate which of several alternative design schemes is likely to meet a specified reliability requirement most economically. Conversely, reliability assessment can also demonstrate which parts of a design scheme are not critical to reliability performance and, therefore, can be made to less stringent requirements without compromising overall reliability and safety.

As an example, an economic assessment may be required of the total Life Cycle Costs for various alternative schemes for a pumping sub-system, for example, two 50 per cent pumps (both operating), or three 50 per cent pumps (either all operating or two operating, one standby), or two 100 per cent pumps (one operating, one standby). Calculations of the likely availability of each of the pumping configurations may be used to estimate the cost of the lost plant output due to pump failures and to investigate the effects on maintenance effort and costs of spares holding requirements. The capital costs of the different pumping configurations can then be weighed against their likely operating and maintenance costs. If the cost of lost plant output is high, as is usually the case for high capital cost, continuously operating, process plants, then the additional capital cost of the standby pumps can easily be justified by the value of plant output saved by the increase in overall plant availability.

(b) Safety

For a safety assessment the most important parameter is usually the likelihood of certain unwanted events occurring on a plant in a given time. The assessment will require the discovery of all potential combinations of events which could result in dangerous failure and the calculation of the probability of their occurrence. Examples of safety assessments are:

(1) calculation of the probability that an aircraft could fail to make an Atlantic crossing;

(2) calculation of the probability that a quantity of poisonous or explosive gas could escape from a

chemical plant at any time during the life of the plant.

In each case a time is involved, whether a few hours or many years; and in each case there is the possibility of loss of life or significant damage to property. With this type of problem a reliability assessment can not only demonstrate the efficacy of proposed safety systems, and procedures, but can also help to determine the cost of increasing safety by assessing the number of additional back-up systems needed to achieve a higher level of safety.

(c) Project viability

Perhaps the most difficult task facing the Project Manager on a large process plant project, apart from any technical problems, is that of convincing the sanctioning authority (whether that be a company main board, or a nationalised public service board) that the project is viable, that is, that the project is not only technically feasible, but that, when completed, the plant will operate safely and reliably with a sufficiently high level of availability to provide a consistent economic return on the capital employed. Here a quantitative reliability assessment can provide the background necessary for rational decisions to be made, instead of those cumulative 'gut feelings' of the engineering project team which have only too often led so many companies to the brink of bankruptcy.

8.3 When to carry out reliability assessments

Whenever possible reliability assessment should be carried out at the initial conceptual design stage (process or engineering flowsheet stage) for a new plant. At this stage, when alternative design schemes are still being considered, it is still possible to influence the significant features of the final design and contribute cost information when choices have to be made between technically acceptable design schemes. It may even provide information which results in the project being cancelled. It therefore becomes a valuable aid to the project manager and his design team.

If an assessment is carried out after the design has been finalised and detailed engineering design is in hand (or even completed), it may be too late, or too costly, to correct any deficiencies or disadvantages highlighted by the assessment.

On existing plants, assessments can be carried out at any time, but it may not be practicable to carry out all desirable modifications because of high costs or the need for a lengthy shutdown. Assessments should be initiated when modifications to the plant are to be carried out, or as part of the investigations of unexpected failure. In the case of plant modifications, the assessment should be carried out as early as possible in the design stage for the reasons stated above for new plant.

8.4 Who should carry out the reliability assessment?

Allocation of responsibility for carrying out the assessment will depend on the nature of the problem and the sort of organization involved. In general terms it should be the responsibility of the project manager in charge of the conceptual design, assisted where appropriate by a reliability specialist. Thus in some cases, for larger user organizations such as ICI, Shell, BP, CEGB, and so on, where the conceptual design is carried out in-house, the assessment would be carried out by the user organization. In other cases, where a plant contractor has been awarded a turnkey contract he would be responsible for the assessment. In cases where large sub-systems are sub-contracted out the sub-contractor could be responsible for an assessment of the sub-systems. For very large process plants a series of reliability assessments may be carried out as the design progresses and changes are made for a multitude of reasons.

8.5 Applications and limitations

The reliability assessment techniques described here in Part Three can be applied to any system from a complete process plant down to plant sub-systems and even to individual pieces of equipment, provided that the system can be divided into sub-sections whose interactions are known, and that something is known about the failure rates and, where appropriate, repair rates of the individual subsections. If reliability data for the sub-sections, whether derived from operating experience with the same or similar plant, or from published data sources, is not available, analysis of system reliability becomes difficult to quantify.

In order to make a clear presentation of the techniques involved in reliability assessment, while keeping the mathematics as simple as possible, it has been assumed here that individual items (or subsections) have constant mean failure rates or constant availabilities. As explained in Part One it is not always correct or appropriate to make this assumption.

A truly constant failure rate is rarely found for an individual item of mechanical equipment during the 'settled down' part of its life because of the presence of wear-out failure mechanisms. However, as discussed in section 3.2, systems which comprise many components which themselves exhibit 'wear-out' effects, eventually (and, usually, very quickly) assume approximately constant failure rates as a result of repair and replacement policies. Thus, for most of the life of a plant, it is usually accurate enough to assume constant failure rate in order to simplify the initial analysis. Then the engineer must use his experience, and knowledge of the plant and equipment, to assess whether he is justified in assuming constant failure rates for the evaluation that he is carrying out.

Apart from simplifying the analysis, another practical reason for the assumption of constant failure rates is that the available data on failures, failure modes, or failure rates is often too limited in quantity or detail to support any other assumption. When extensive data is available which shows that 'reliability growth' or 'wear-out' is occurring, then other reliability analysis methods become appropriate.

When failure rates are found to be nearly constant,

use of the techniques described here in Part Three will produce results which are of the right order of magnitude. If, however, 'wear-out' mechanisms are present, the results are likely to be somewhat optimistic. In practice this may not necessarily matter too much, but (particularly if major technical or cost decisions are to be made on the basis of these results) it is advisable to carry out a sensitivity analysis. This would involve repeating the calculations with different failure rate values (chosen from the range of likely minimum to maximum values) to check the sensitivity of the results to different assumptions.

CHAPTER 9. THE ANALYSIS OF SIMPLE SYSTEMS

9.1 Reliability diagrams

(a) *Reliability block diagrams*

The first stage of the design of complex process or engineering plant is to construct a block diagram (or flow-diagram) in which each block represents one of the plant's constituent sub-systems or items of equipment. Thus, a *schematic* block diagram shows how the items are physically connected, while a *functional* block diagram shows the flow of power, material, etc., through the system, with the relationship between input and output specified for each block. This diagram assists the conceptual design of the system and must be approved before any detailed engineering design can proceed.

Similarly, the assessment of overall system reliability may be facilitated by the construction and analysis of a Reliability Block Diagram (RBD). In a reliability block diagram the connections between the items symbolise the ways in which the system will *function as required* and do not necessarily indicate the actual physical connections. Also, for ease of analysis, it is usual to model each item as being either in the fully working or in the totally failed state (a conservative, or pessimistic, reliability assessment would model partial failure or reduced output of an item as a total failure). A Reliability Block Diagram is usually drawn up using a schematic or functional diagram of the plant as a basis.

As a simple illustration, consider a system consisting of a motor fed by two fuel pumps, both of which are normally working. The system is so designed that, provided the motor is working, full power output will be achieved even if one of the pumps has failed. The Reliability Block Diagram would then be as in Fig. 9.1, which indicates that output may be achieved via either, or both, of the pumps.

Although Fig. 9.1 would probably be identical in form to the relevant schematic or functional diagram (as in this case the most likely physical arrangement would consist of two parallel fuel lines with one pump in each), this need not necessarily be the case. Figure 9.1 could still be the appropriate Reliability Block Diagram even if the pumps were located in series along a single fuel line, *provided that full power output would be achieved even with one pump failed*. The form of the Reliability Block Diagram is determined by the system logic for achieving the required performance, not simply by the system layout.

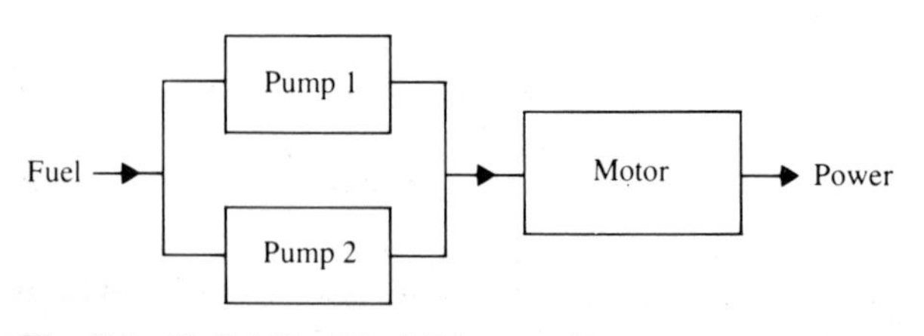

Fig. 9.1 Reliability Block Diagram for pump-motor system

(b) *Reliability logic diagrams*

Another type of block diagram useful for reliability analysis is the Reliability Logic Diagram (RLD). Although this is also usually constructed from the information revealed by a schematic or functional diagram, it bears little resemblance, in its purest form, to either of these. It represents combinations of outputs, alternative paths, etc., by the use of logic gates rather than symbolic flow lines (this type of logic diagram is also extensively used in computer programming). Figure 9.2(a) is the Reliability Logic Diagram for the pump-motor system drawn in 'success' notation (that is, where each block indicates an item in the working state and where the output is the achievement of system requirement). Figure 9.2(b) shows the complementary Reliability Logic Diagram in 'failure' notation (where each block indicates an item in the not working or failed state) and is essentially a simplified fault tree (see Chapter 13). It is fairly common practice to construct reliability diagrams which use a mixture of both Reliability Block Diagrams and Reliability Logic Diagrams to represent system logic.

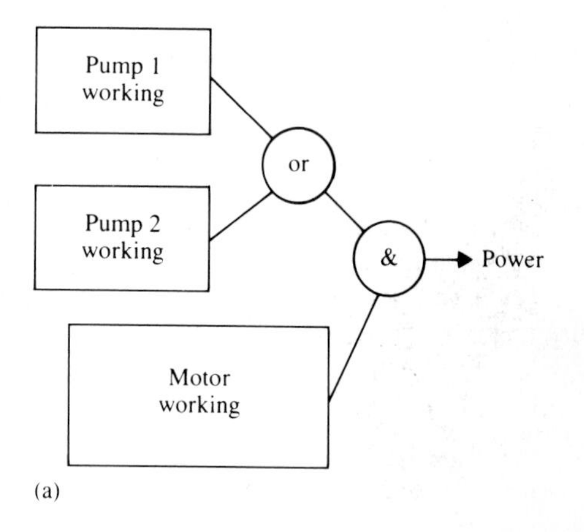

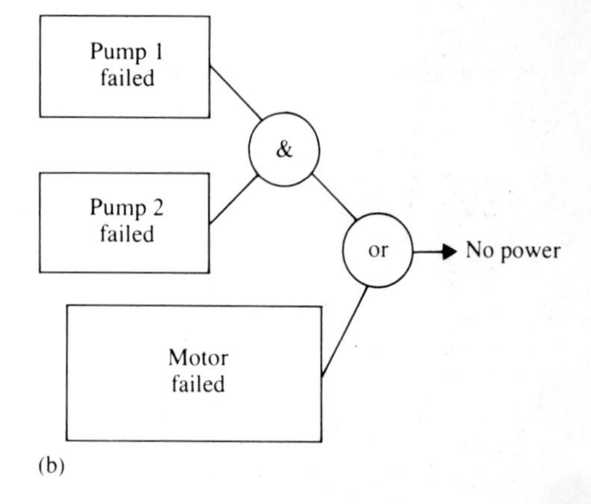

Fig. 9.2. Reliability Logic Diagram for pump-motor system: (a) in 'success' notation, (b) in 'failure' notation

9.2 Analysis of simple systems

Series and active parallel Reliability Block Diagrams are the simplest building blocks of reliability analysis. Techniques for analysing series reliability block diagrams are described in section 9.3, and techniques for analysing active parallel Reliability Block Diagrams are described in section 9.4. Reliability Block Diagrams of complex systems can often be analysed by the application of simple series and active parallel system analysis, as described in Chapter 15.

The approach used here to analyse the diagrams is to first lay out the basic methods in terms of probability. This presents the methods in their simplest mathematical form and gives the general case which can be readily extended to cover the particular cases of reliability and availability.

By taking a Reliability Block Diagram, for example, Fig. 9.1, and assigning to each block a probability that it is in the working state, the probability that the whole system is in the working state can be found.

Probability and reliability are not interchangeable. A probability is a dimensionless quantity and, therefore, independent of time. In contrast, a reliability is the probability that an item or system survives for a period of time, and so is time dependent. Confusion of the two can lead to calculation errors. Reliability must also not be confused with availability. Reliability is a measure of the time that a system will work without failure or repair, whereas availability is a measure of the percentage of time that a system is working over a long period during which it can fail and be repaired several times.

When using probability techniques it is customary to define the degree of probability as a decimal fraction of unity. Thus, a probability of one means absolute certainty and a probability of zero indicates absolute impossibility. For an item which can be only in either a working or a failed state, the sum of the probability that it is working (P) and the probability that it has failed (F) is unity. That is

$$P + F = 1 \qquad (9.1)$$

Probability and availabilities may also be quoted as percentages.

9.3 Analysis of simple series systems

(a) Theory

If a system is made up of two units and for system success both must work, then for a reliability assessment the units are considered to be in series. Figure 9.3 shows the appropriate Reliability Block Diagram and also shows, for comparison, the equivalent Reliability Logic Diagram in 'success' notation. As emphasised earlier, such diagrams say nothing about the actual physical connection which could be in series (for example, pump followed by non-return valve) or in parallel (for example, fuel and ignition sub-systems of an internal combustion engine).

The assumption that failure of either unit occurs quite independently of failure of the other will be adopted. This will simplify the analysis and may often be valid (or at least a plausible first approximation). The probability, P_s, that the system is working, that is, that units 1 and 2 are both working, is then given by the product of the two separate probabilities, P_1, that unit 1 is working, and P_2, that unit 2 is working. That is

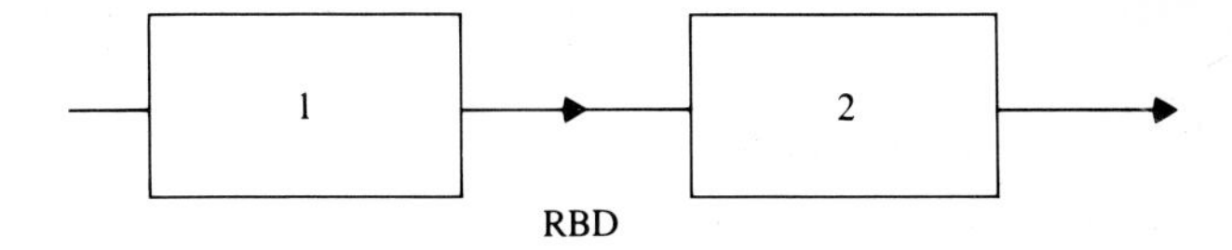

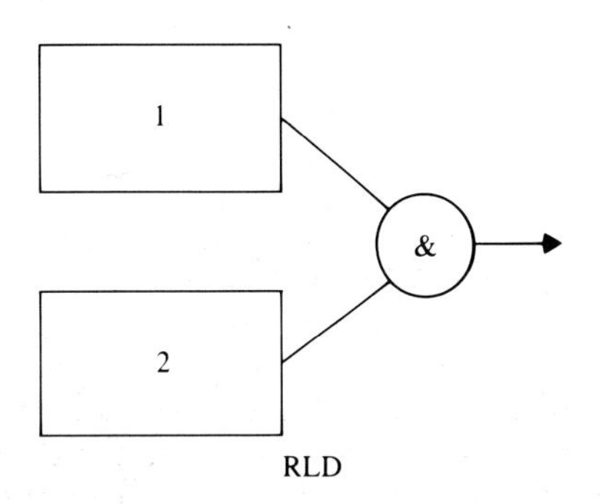

Fig. 9.3. Reliability Block Diagram and Reliability Logic Diagram ('success' notation) for two units in series

$$P_s = P_1 \times P_2 \qquad (9.2)$$

If, for the individual units or for the system, only two states, working or failed, are possible then the probability, F_s, that the system is in the failed state is (from equations 9.1 and 9.2)

$$F_s = 1 - P_s = 1 - (P_1 \times P_2)$$
$$= 1 - (1 - F_1)(1 - F_2)$$

or

$$F_s = F_1 + F_2 - (F_1 \times F_2) \qquad (9.3)$$

where F_1 is the probability that unit 1 will be in the failed state and F_2 is the probability that unit 2 will be in the failed state. For a system with more than two units in series, as shown in Fig. 9.4, the probability of system success is obtained by simple extension of equation (9.2); that is

$$P_s = P_1 \times P_2 \times P_3 \times \cdots \times P_n \qquad (9.4)$$

and hence the probability of system failure is

$$F_s = 1 - (P_1 \times P_2 \times \cdots \times P_n)$$
$$= 1 - (1 - F_1)(1 - F_2) \cdots (1 - F_n) \qquad (9.5)$$

(b) Series system reliability – applications and examples

(i) Availability of a series system

Availability, A, was defined in section 3.3 and equation (3.1). In the long term, the availability of a unit which is

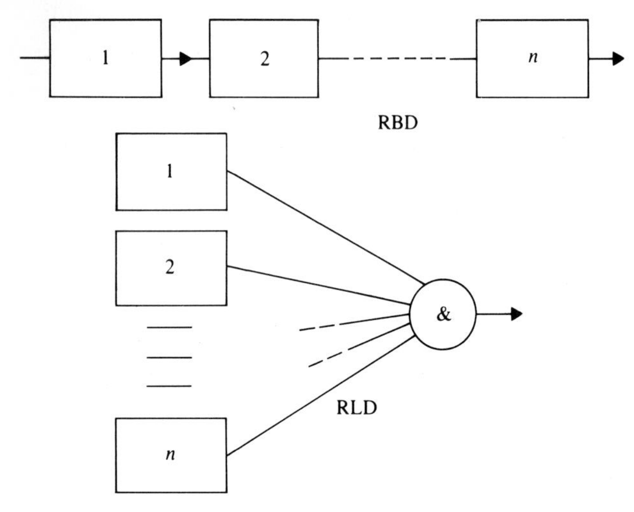

Fig. 9.4. **Reliability Block Diagram and Reliability Logic Diagram ('success' notation) for a system with *n* units in series**

subject to randomly occurring 'failure and repair' forced outages is equivalent to the probability (or proportion of time) that it is in the working state. Conversely, unavailability, U, is the proportion of time that a unit is in the failed state; that is

$$A + U = 1 \tag{9.6}$$

Consider, for example, a simple system which is required to run continuously, has no redundant* units (that is, the units are in series), and in which the units are subject to randomly occurring 'failure and repair' forced outages which result in average unit availabilities A_1, A_2, and so on (hence average unit unavailabilities, $U_1 = 1 - A_1$, $U_2 = 1 - A_2$, and so on). The system is available only when all the units are working, so over a long period during which each unit is repaired when it fails, from equations (9.4) and (9.5), for a system with n units in series, the system availability, A_s, and unavailability, U_s, are given by

$$A_s = A_1 \times A_2 \times A_3 \times \cdots \times A_n \tag{9.7}$$

$$U_s = 1 - (A_1 \times A_2 \times \cdots \times A_n) \tag{9.8}$$

or

$$U_s = 1 - (1 - U_1)(1 - U_2) \cdots (1 - U_n) \tag{9.9}$$

If unit unavailabilities are very low, i.e., each $U \ll 1$, then equation 9.9 may be simplified with sufficient accuracy, to

$$U_s \simeq U_1 + U_2 + U_3 + \cdots + U_n \tag{9.10}$$

provided that U_s is not more than approximately 0.1.

Example 9.1

For the system in Fig. 9.5 we wish to find the system availability and, hence, its unavailability.

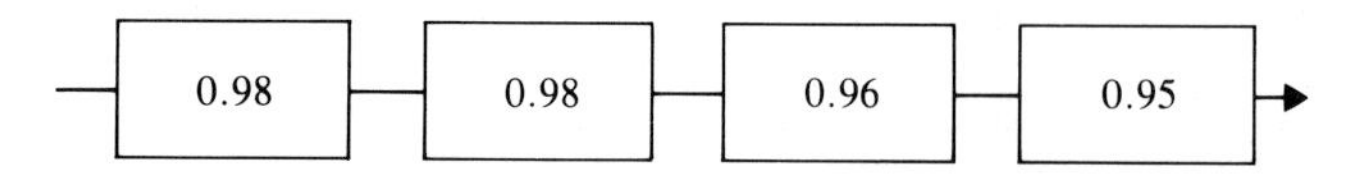

Fig. 9.5. **Reliability Block Diagram for a series system of four units. Unit availabilities are as shown**

From equations (9.6) and (9.7)

$$A_s = 0.98 \times 0.98 \times 0.96 \times 0.95$$
$$= 0.88$$

$$U_s = 1 - A_s$$
$$= 1 - 0.88$$
$$= 0.12$$

Since, in this example, the unit unavailabilities are all $\ll 1$ (for example, $U_1 = 1 - A_1 = 1 - 0.98 = 0.02$), equation (9.10) could have been used with little error to obtain the system unavailability; that is

$$U_s \simeq 0.02 + 0.02 + 0.04 + 0.05$$
$$\simeq 0.13$$

The result for system availability illustrates the general rule that the reliability (and consequently availability) of a series system arrangement is inevitably less than that of its least reliable (available) constituent. A good example of this is a packing line for pharmaceutical, cosmetic, or food products, where it is not unusual to have up to eight units in series (i.e., sorter, filler, capper, labeller, code render, cartonner, carton compiler, overwrapper). If it is assumed that the availability of each constituent unit is equal, then a graph of system availability against unit availability can be drawn for a series system of n units and this is shown in Fig. 9.6.

From Fig. 9.6 it can be seen that for a series system of many units the average availability of the units must be high if an acceptable level of system availability is to be achieved.

(ii) Reliability of a series system

If a system is required to run for a given time, t, and its constituent units exhibit constant mean failure rates λ_1,

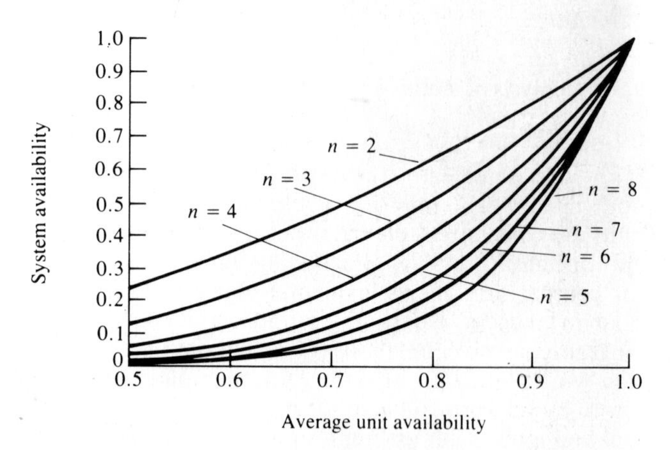

Fig. 9.6. **System availability for a series system of *n* units with identical availabilities**

---

* In a reliability context 'redundant' means installed spare items of plant or equipment.

$\lambda_2, \ldots \lambda_n$, then their respective probabilities of *not* failing prior to t (their reliabilities or *survival probabilities*) are given by (see Chapter 6)

$$R_1(t) = e^{-\lambda_1 t}, \quad R_2(t) = e^{-\lambda_2 t}, \ldots \text{ and}$$
$$R_n(t) = e^{-\lambda_n t},$$

and the overall system reliability for the time, t, is derived using equation (9.4). That is

System reliability,

$$\begin{aligned} R_s(t) &= R_1(t) \times R_2(t) \times \cdots \times R_n(t) \\ &= e^{-\lambda_1 t} \times e^{-\lambda_2 t} \times \cdots \times e^{-\lambda_n t} \\ &= e^{-(\lambda_1+\lambda_2+\cdots+\lambda_n)t} \end{aligned} \qquad (9.11)$$

Thus, for the series system the expression for its reliability (*survival probability*) is of the same form, $e^{-\text{const.}t}$, as that for a single unit, the system behaving like a single unit with a constant mean failure rate, λ_s, which is equal to the sum of the failure rates of the system's constituent units. That is

$$\lambda_s = \lambda_1 + \lambda_2 + \cdots + \lambda_n$$

The probability of system failure prior to time t is found by comparison with equation (9.5)

$$F_s(t) = 1 - e^{-(\lambda_1+\lambda_2+\cdots+\lambda_n)t} \qquad (9.12)$$

From the above results it can be shown that the Mean Time To Failure (MTTF), or average life, of the system is

$$(\text{MTTF})_s = 1/(\lambda_1 + \lambda_2 + \cdots + \lambda_n) \qquad (9.13)$$

If the mean lives, $1/\lambda$, of the individual units are at least an order of magnitude greater than the required survival time, t, that is, $\lambda t \ll 1$, then

$$e^{-\lambda t} \simeq 1 - \lambda t$$

and

$$R_s(t) \simeq 1 - (\lambda_1 + \lambda_2 + \cdots + \lambda_n)t \qquad (9.14)$$

$$F_s(t) \simeq (\lambda_1 + \lambda_2 + \cdots + \lambda_n)t \qquad (9.15)$$

This is a convenient simplification as long as it is only used when $\lambda t \ll 1$. Failure to observe this condition will lead to significant calculation errors.

Example 9.2

Units 1, 2, and 3 of Fig. 9.7 are identical pumps, each exhibiting a constant mean failure rate of 10^{-5}/hour, that is, the MTTF of each pump $= 1/10^{-5} = 10^5$ hour. All three must be working for system success. Calculate (a) the system MTTF, (b) the system reliability (survival probability) at 10^4 hours, (c) the system reliability at 10^3 hours.

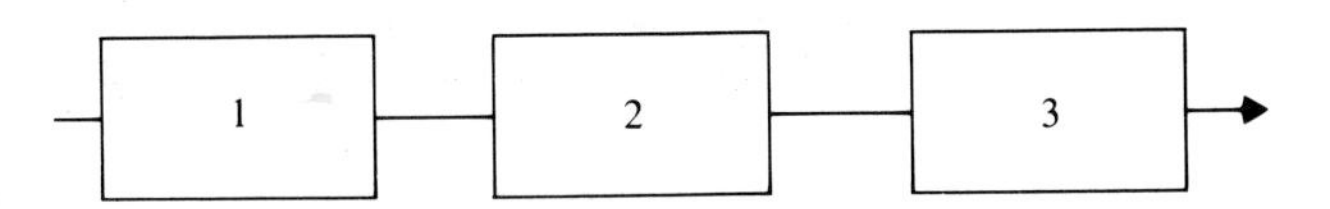

Fig. 9.7. Reliability Block Diagram of a series system with identical units

(a) From equation (9.13)

$$\begin{aligned} \text{System MTTF} &= 1/(10^{-5} + 10^{-5} + 10^{-5}) \\ &= 3.3 \times 10^4 \text{ hours} \end{aligned}$$

(b) If the required system survival time is 10^4 hours then from equation (9.11)

$$\begin{aligned} \text{System reliability} &= R_s(10^4 \text{ hours}) \\ &= e^{-(10^{-5}+10^{-5}+10^{-5})10^4} \\ &= e^{-0.3} \\ &= 0.74 \end{aligned}$$

(c) If the required survival time is 10^3 hours then, for each unit,

$$\begin{aligned} \lambda t &= 10^{-5} \times 10^3 \\ &= 10^{-2} \end{aligned}$$

that is

$$\lambda t \ll 1$$

and, therefore, equation (9.14) applies; that is

System reliability,

$$\begin{aligned} R_s(t) &\simeq 1 - (\lambda_1 + \lambda_2 + \cdots + \lambda_n)t \\ R_s(10^3) &\simeq 1 - (10^{-5} + 10^{-5} + 10^{-5})10^3 \\ &\simeq 1 - (3 \times 10^{-2}) \\ &\simeq 0.97 \end{aligned}$$

9.4 Analysis of simple active parallel systems

(a) Theory

If a system consists of two units, both of which are normally working, that is, contributing to the system's operation, but the system requirement will still be met even if one is failed, then from a reliability point of view, they are considered to be in parallel. The Reliability Block Diagram of such a *fully-redundant system* is shown in Fig. 9.8, which also shows for comparison, the equivalent Reliability Logic Diagram in success notation.

In section 9.3(a) it was shown (see Fig. 9.3) that, with two units in series, system *success* requires both units to *succeed*. Here, with two units in an active parallel system, system *failure* requires both units to *fail*, that is, the reliability logic is reversed. If, as before, we assume that unit failures are statistically independent, then, for the

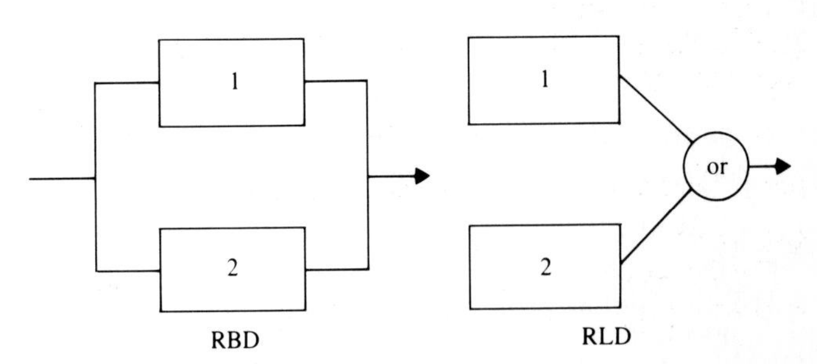

Fig. 9.8. Reliability Block Diagram and Reliability Logic Diagram ('success' notation) for two units in active parallel system

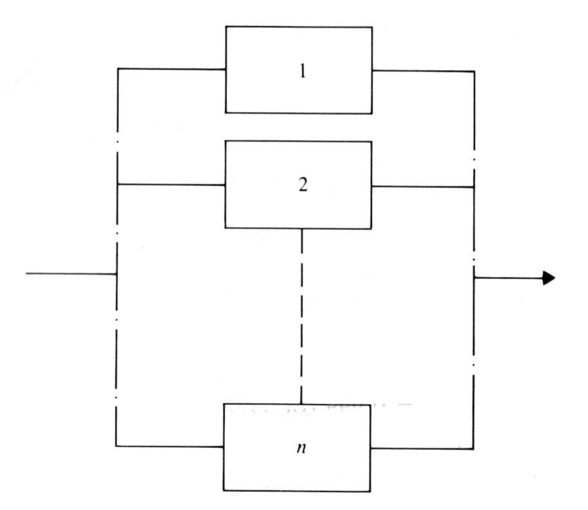

Fig. 9.9. Reliability Block Diagram for active parallel system with n units (system success if at least one unit working; system failure if all units failed)

simple active parallel system, the probability, F_s, that the system is in the failed state is

$$F_s = F_1 \times F_2 \tag{9.16}$$

Thus, the probability, P_s, that the system is working is

$$P_s = 1 - F_s = 1 - (F_1 \times F_2)$$
$$= 1 - (1 - P_1)(1 - P_2)$$

or

$$P_s = P_1 + P_2 - P_1P_2 \tag{9.17}$$

This analysis can be extended to cover systems with more than two units in active parallel, in which all units are normally working but the system succeeds if at least one unit is working, that is, 1 out of n.

The Reliability Block Diagram is as in Fig. 9.9, and the probability, F_s, that the system is in the failed state is obtained by simple extension of equation (9.16). That is

$$F_s = F_1 \times F_2 \times F_3 \times \cdots \times F_n \tag{9.18}$$

and, hence

$$P_s = 1 - F_1 \times F_2 \times \cdots \times F_n$$
$$= 1 - (1 - P_1)(1 - P_2) \cdots (1 - P_n) \tag{9.19}$$

The general case has been shown, although, in practice, it is unusual to find more than three or four items in active parallel.

(b) Simple active parallel system reliability: applications and examples

(i) Availability of active parallel system

A system, required to run continuously, has units subject to random outages giving average unit availabilities, A_1, A_2, and so on, and hence unavailabilities $U_1 = 1 - A_1$, $U_2 = 1 - A_2$, and so on.

The system is available if at least one unit is available. By comparison with equations (9.18) and (9.19), system availability, A_s, and unavailability, U_s, are given by

$$U_s = U_1 \times U_2 \times \cdots \times U_n \tag{9.20}$$

$$A_s = 1 - (U_1 \times U_2 \times \cdots \times U_n)$$
$$= 1 - (1 - A_1)(1 - A_2) \cdots (1 - A_n) \tag{9.21}$$

Example 9.3

For the system in Fig. 9. 10 we wish to find the system availability and unavailability.

Unit availabilities are as follows

$$A_1 = 0.60, \quad A_2 = 0.70, \quad A_3 = 0.80$$

Unit unavailabilities are, therefore

$$U_1 = 0.40, \quad U_2 = 0.30, \quad U_3 = 0.20$$

For the system, equations (9.20) and (9.21) give

$$U_s = 0.40 \times 0.30 \times 0.20$$
$$= 0.024 \text{ (or 2.4 per cent)}$$

$$A_s = 1 - 0.024$$
$$= 0.976 \text{ (or 97.6 per cent)}$$

This result illustrates what commonsense would suggest, that the availability of a redundant system will always be greater than that of its components.

(ii) Reliability of an active parallel system

A system is required to run for a given time, t, and its units exhibit mean failure rates, $\lambda_1, \lambda_2, \ldots \lambda_n$. As it is an active parallel system it survives if at least one unit survives.

From equation (9.18) the probability of the system failing prior to time t is

$$F_s(t) = F_1(t) \times F_2(t) \times \cdots \times F_n(t)$$
$$= (1 - e^{-\lambda_1 t})(1 - e^{-\lambda_2 t}) \cdots (1 - e^{-\lambda_n t}) \tag{9.22}$$

(For a system with n *identical* units $F_s(t) = (1 - e^{-(\lambda t)n})$, hence, system reliability is given by

$$R_s(t) = 1 - F_s(t)$$
$$= 1 - (1 - e^{-\lambda_1 t})(1 - e^{-\lambda_2 t}) \cdots (1 - e^{-\lambda_n t}) \tag{9.23}$$

From the above result an expression for the system mean

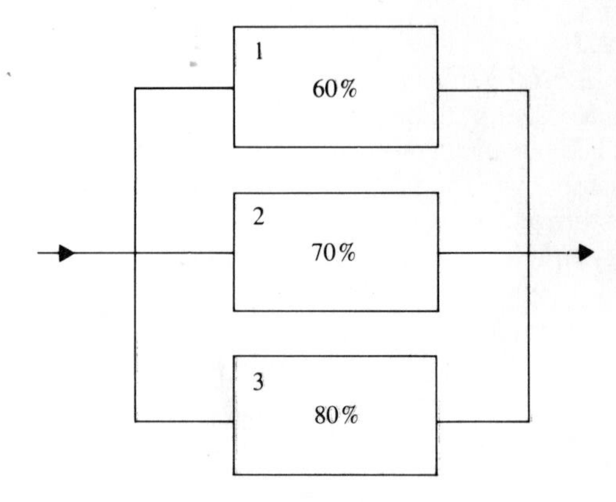

Fig. 9.10. Availability Block Diagram for active parallel system of three chemical reactors (system failed only if all three reactors failed; reactor availabilities are as shown)

time to failure can be found. However, its form is more complex than that for a series system. For an active parallel system comprising two units

$$\text{MTTF} = \frac{1}{\lambda_1} + \frac{1}{\lambda_2} - \frac{1}{\lambda_1 + \lambda_2} \qquad (9.24)$$

For an active parallel system comprising three units

$$\text{MTTF} = \frac{1}{\lambda_1} + \frac{1}{\lambda_2} + \frac{1}{\lambda_3} - \frac{1}{(\lambda_1 + \lambda_2)} - \frac{1}{(\lambda_1 + \lambda_3)} - \frac{1}{(\lambda_2 + \lambda_3)} + \frac{1}{(\lambda_1 + \lambda_2 + \lambda_3)} \qquad (9.25)$$

If the unit mean lives are more than an order of magnitude greater than the mission time, that is, $\lambda t \ll 1$, then

$$e^{-\lambda t} \simeq 1 - \lambda t$$

and

$$F_s(t) \simeq (\lambda_1 \times \lambda_2 \times \cdots \times \lambda_n) \times t^n \qquad (9.26)$$

Example 9.4

Three identical vacuum pumps are normally all working, but as long as at least one is working then an adequate vacuum will be maintained. Pumps of this type have been found to exhibit a constant mean failure rate of 10^{-5}/hr. Calculate (a) system MTTF, (b) system reliability for a run of 10^4 hours, (c) system reliability for a run of 10^3 hours.

(a) From equation (9.25)

$$\text{System MTTF} = \frac{3}{10^{-5}} - \frac{3}{2 \times 10^{-5}} + \frac{1}{3 \times 10^{-5}}$$
$$= 1.8 \times 10^5 \text{ hours}$$

(b) Run time 10^4 hours

$$\lambda t = 10^{-5} \times 10^4 = 0.1$$

From equation (9.22)

$$F_s(10^4 \text{ hrs}) = (1 - e^{-0.1})^3$$
$$= 8.6 \times 10^{-4} \text{ (or } 10^{-3} \text{ approx.)}$$

hence, from equation (9.23)

$$R_s(10^4 \text{ hrs}) \simeq 1 - 10^{-3}$$
$$= 0.999 \text{ (or 99.9 per cent)}$$

(c) Run time 10^3 hours

$$\lambda t = 10^{-5} \times 10^3 = 0.01$$

Equation (9.26) can be used since $\lambda t \ll 1$

$$F_s(10^3 \text{ hrs}) \simeq (10^{-5} \times 10^3)^3$$
$$\simeq 10^{-6}$$

hence

$$R_s(10^3 \text{ hrs}) \simeq 1 - 10^{-6}$$
$$\approx 1$$

CHAPTER 10. ACTIVE PARALLEL SYSTEMS WITH PARTIAL REDUNDANCY AND SYSTEMS WITH STANDBY UNITS

Chapter 9 has dealt with simple active parallel systems, that is, simple parallel systems in which all units are normally running, but only one unit needs to be running for system success. However, not all parallel systems are included in this description. In this chapter the analysis is extended to cover three further groups of systems.

(1) Parallel systems in which all units are normally working but *more than one* unit must be running for system success; these are known as *Active Parallel Systems with Partial Redundancy*.

(2) Standby systems in which all the units required for system success are running but a number of spares are held in readiness to replace any failed unit immediately; these are known as *Standby Systems* or *Inactive Parallel Systems*.

(3) Dormant systems in which no units are normally running but at least one unit must start on demand, usually to protect plant in abnormal operating conditions.

10.1 Active parallel systems with partial redundancy

In this section an active parallel system, which has several identical units, is considered. The system is successful if at least *some* of the units are working. Such a system might be an aircraft with four completely independent engines which will fly safely provided that at least any two engines are working.

The general case is shown in Fig. 10.1. Each unit is identical and has a probability, F, of being in the failed state ('down'), and a probability, P, of being in the working state ('up'). As before, it will be assumed that these are the only states possible (that is, partial failures will not be modelled), so (from equation (9.1))

$$P + F = 1.$$

Also it will be assumed that unit failures are statistically independent, and that the system runs successfully if at least r units are working.

For this general case it can be shown that the probability that any particular number, r, of units will be working (up) can be expressed as a term in the expansion of the binomial expression $(F + P)^n$, where n is the total number of units, as shown in Table 10.1.

As system success has been defined as having at least r units working, then those states below the line in Table 10.1 are those which result in system success.

Therefore the total probability of system success, P_S, is the sum of the probabilities of *all* the states *below* the line, that is

$$P_S = {}^nC_r P^r F^{n-r} + {}^nC_{r+1} P^{r+1} F^{n-r-1} + \cdots + P^n$$

$$= \sum_{m=r}^{m=n} {}^nC_m P^m F^{n-m} \qquad (10.1)$$

The total probability of system failure, F_S, is

$$F_S = 1 - \sum_{m=r}^{m=n} {}^nC_m P^m F^{n-m} = \sum_{m=0}^{m=r-1} {}^nC_m P^m F^{n-m} \qquad (10.2)$$

that is, the sum of the probability of *all* the states *above* the line in Table 10.1

Table 10.1. Probability that any number, r, out of n units is working (up)

State	*Probability*	*Comment*
0 units up	F^n $(= {}^nC_0 P^0 F^n)$	where ${}^nC_0 = 1, P^0 = 1$
1 unit up	${}^nC_1 P F^{n-1}$	
2 units up	${}^nC_2 P^2 F^{n-2}$	
$(r-1)$ units up	${}^nC_{r-1} P^{r-1} F^{n-r+1}$	
r units up	${}^nC_r P^r F^{n-r}$	
n units up	P^n $(= {}^nC_n P^n F^0)$	where ${}^nC_n = 1, F^0 = 1$

${}^nC_r$ *is the number of ways (combinations) of selecting* r units out of a total of n units, irrespective of the order in which they are selected,

$${}^nC_r = \frac{n!}{r!\,(n-r)!},$$

where

$$n! = n \text{ factorial} = n\cdot(n-1)\cdot(n-2)\cdots(2)\cdot(1)$$

Thus

$$6! = 6 \times 5 \times 4 \times 3 \times 2 \times 1$$

and

$${}^4C_2 = \frac{4!}{2!\,(4-2)!} = \frac{4 \times 3 \times 2 \times 1}{(2 \times 1)(2 \times 1)} = 6$$

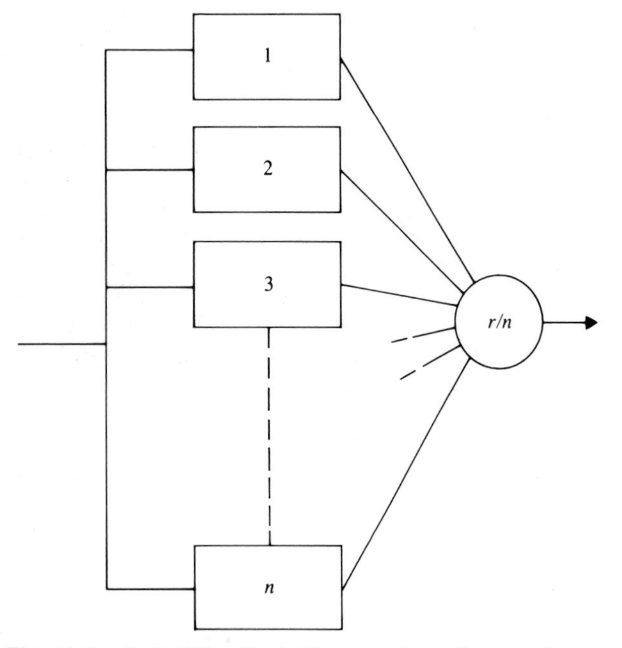

Fig. 10.1. Reliability Block Diagram for active parallel system with n identical units, at least r of which need to be working for system success

Table 10.2. Probabilities of success and failure for commonly encountered partially redundant systems

System	*Probability of success,* P_S	*Probability of failure,* F_S
2 out of 3	$P^3 + 3P^2F$	$F^3 + 3PF^2$
2 out of 4	$P^4 + 4P^3F + 6P^2F^2$	$F^4 + 4PF^3$
3 out of 4	$P^4 + 4P^3F$	$F^4 + 4PF^3 + 6P^2F^2$

The probability of system success for any pattern of active parallel units with partial redundancy may be found by using the equation for the general case, that is, equation (10.1). While this, and equation (10.2) for system failure, may look daunting for those not familiar with this branch of mathematics, in practice the most commonly encountered cases are 2 out of 3, 2 out of 4, and 3 out of 4. For these specific cases the probabilities of system success and system failure are given in Table 10.2.

(a) Reliability of active parallel systems with partial redundancy: applications and examples

(i) Application to availability

Consider a system which is required to run continuously and has n identical units subject to random outages giving an average unit availability of A. The system succeeds if at least r units are available.

The system availability can be deduced from equation (10.1)

$$A_S = \sum_{m=r}^{m=n} {}^nC_m A^m (1 - A)^{n-m} \quad (10.3)$$

Example 10.1

Find the availability of the system shown in Fig. 10.2.

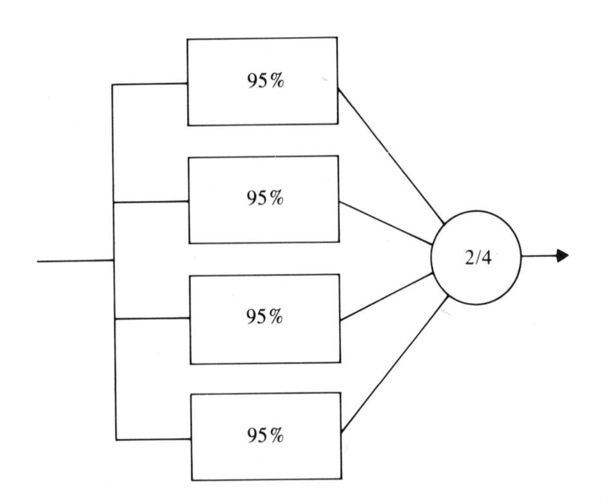

Fig. 10.2. Availability Block Diagram for four identical heat exchangers: for system success at least two exchangers must be working. (The availability of each heat exchanger is 0.95.)

From equation (10.3) the availability of such a '2 out of 4' system is

$$A_S = \sum_{m=2}^{4} {}^4C_m 0.95^m (1 - 0.95)^{4-m}$$

which (by similarity with the equation for success of a '2 out of 4' system on Table 10.2) can be restated as

$$\begin{aligned} A_S &= A^4 + 4A^3(1 - A) + 6A^2(1 - A)^2 \\ &= 0.95^4 + 4 \times 0.95^3 \times 0.05 + 6 \times 0.95^2 \times 0.05^2 \\ &= 0.8145 + 4 \times 0.8574 \times 0.05 + 6 \times 0.9025 \\ &\quad \times 0.0025 \\ &= 0.8145 + 0.1715 + 0.0135 \\ &= 0.9995 \text{ (or 99.95 per cent)} \end{aligned}$$

(ii) Application to system reliability

Consider a system which is required to run for a given time, t, and has n units, each exhibiting a constant mean failure rate, λ, so that the reliability of a given unit at time t is given by

$$R(t) = e^{-\lambda t}$$

Assume that the system survives if at least r units survive. From equation (10.1) the system reliability at time t is

$$R_S(t) = \sum_{m=r}^{n} {}^nC_m (e^{-\lambda t})^m (1 - e^{-\lambda t})^{n-m} \quad (10.4)$$

and from equation (10.2) the probability of system failure short of the mission time (that is, before time t) is

$$F_S(t) = \sum_{m=0}^{r-1} {}^nC_m (e^{-\lambda t})^m (1 - e^{-\lambda t})^{n-m} \quad (10.5)$$

From the above an analytical expression for the system Mean Time To Failure, MTTF, *can* be derived. However, apart from the simplest cases, such as active parallel systems with 2 or 3 units (see equations (9.24) and (9.25)), these expressions would be usually rather unwieldy, and it is not appropriate to develop this further at this stage.

Example 10.2

An aircraft has four completely independent identical engines, but will continue to fly safely even if two have failed. The observed constant mean failure rate per engine (given that it has started successfully) is 10^{-2}/hr (that is the MTTF per engine is 100 hours). Given that the aircraft has taken off successfully, what is the probability that it will *not* fly safely for 10 hours (assuming that all failures other than engine failure are negligible)?

For each individual engine

$$R(10 \text{ hrs}) = e^{-\lambda t} = e^{-0.01 \times 10} = e^{-0.1} = 0.905$$

$$F(10 \text{ hrs}) = 1 - e^{-\lambda t} = 1 - 0.905 = 0.095$$

For the aircraft, probability of failure in under 10 hours is found from equation (10.5); that is

$$F_S(10 \text{ hrs}) = \sum_{m=0}^{1} {}^4C_m \times 0.905^m \times 0.095^{4-m}$$

which (by comparison with Table 10.2 for a '2 out of 4' system) can be restated as

$$F_S(10 \text{ hrs}) = 0.095^4 + 4 \times 0.905 \times 0.095^3$$

(Probability of all 4 engines failing) (Probability of any 3 engines failing)

$$= 8.15 \times 10^{-5} + 3.10 \times 10^{-3}$$
$$= 3.18 \times 10^{-3} \text{ (0.318 per cent)}$$

10.2 Standby or inactive-parallel systems

(a) Introduction

Chapter 9 illustrated how the arrangement of units in active parallel will improve reliability. In many, if not most, mechanical systems, however, this may not be desirable or even feasible. For example, it might well be that if a vessel were to be fed by two positive-displacement pumps in active parallel it would be over-pressurized. An alternative would be to have one pump working and the other on standby, ready to take over should the normally working pump fail, as indicated in Fig. 10.3.

The evaluation of the reliability of such a system presents rather more difficulty than the systems examined so far, even with the adoption of simplifying modelling assumptions such as statistical independence of unit failures, no repair, negligible flow-line failures, and so on. Even then there are several ways in which the system could fail.

- failure of pump 1, successful changeover, eventual failure of pump 2 (or pump 2 already failed while on standby);
- failure of pump 1, successful changeover, eventual spurious changeover back to failed pump 1;
- failure of pump 1, failure of changeover unit to respond;
- failure of changeover unit to transmit flow.

If the system reliability modelling includes the possibility of repair of the off-line unit then analysis can be very complex, possibly taking into account such factors as repair time, repair-team availability, priority rules, and so on. Only the simpler models will be considered here, and results only are quoted.

(b) Standby systems with perfect changeover

Consider a system which has one operating unit and one, or more, identical standby units, and assume that no repairs are carried out on failed units. Such a system is depicted in Fig. 10.4.

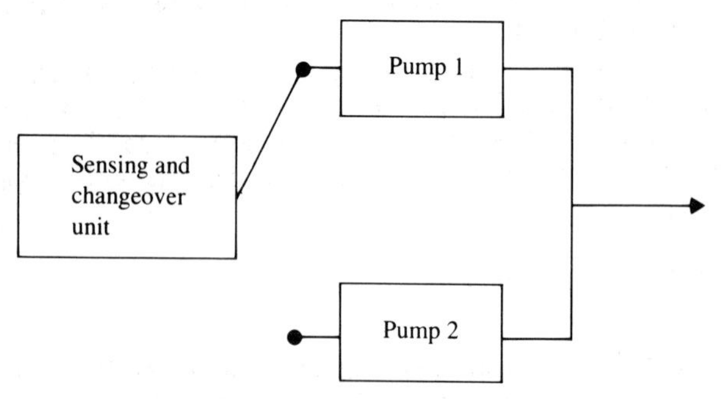

Fig. 10.3. Standby system

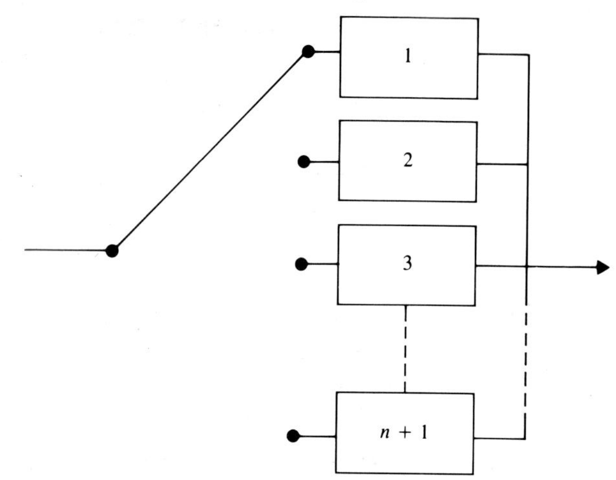

Fig. 10.4. System with one operating unit and one or more identical standby units: failed units not repaired

The reliability (survival probability), $R_S(t)$, and Mean Time To Failure (MTTF) of the system are given in Table 10.3.

$R_S(t)$ is the system reliability (the probability that the system will survive for the time t) and λ is the average failure rate per unit. Unit 1 operates until it fails, then unit 2 takes over, and so on, until the last standby has failed, when the system also fails. The changeover unit is assumed not to fail.

(c) Standby systems with imperfect changeover

Consider a system similar to that described in the previous section, but which also has the possibility that the changeover unit, which brings the standby units on line when the operating unit fails, might not work correctly on demand (that is, it is imperfect). Such a system is depicted in Fig. 10.5.

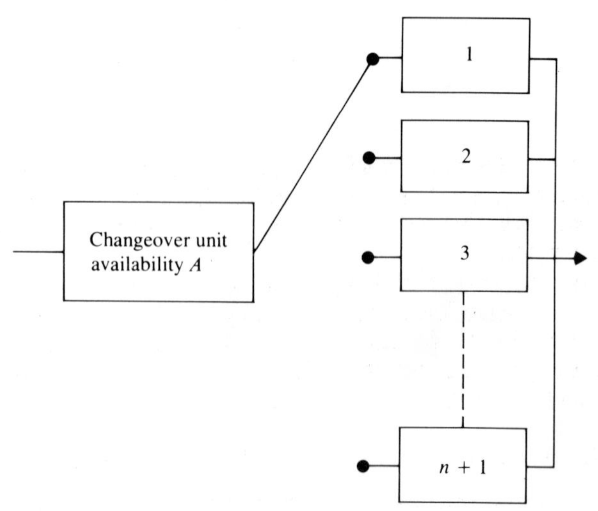

Fig. 10.5. System with one operating unit and one or more identical standby units with imperfect changeover unit: failed units not repaired

Table 10.3. Reliability and MTTF for system with one operating unit and *n* standby units

No. of standbys	*System reliability (survival probability)*	*MTTF*
1	$R_S(t) = e^{-\lambda t}(1 + \lambda t)$	$\frac{2}{\lambda}$
2	$R_S(t) = e^{-\lambda t}\left\{1 + \lambda t + \frac{(\lambda t)^2}{2!}\right\}$	$\frac{3}{\lambda}$
n	$R_S(t) = e^{-\lambda t}\left\{1 + \lambda t + \frac{(\lambda t)^2}{2!} + \cdots + \frac{(\lambda t)^n}{n!}\right\}$	$\frac{1+n}{\lambda}$

Table 10.4. Reliability and MTTF for system with one operating unit, *n* standby units, and imperfect changeover

No. of standbys	*System reliability (survival probability)*	*MTTF*
1	$P_S(t) = e^{-\lambda t}(1 + A\lambda t)$	$\frac{1+A}{\lambda}$
2	$P_S(t) = e^{-\lambda t}\left\{1 + A\lambda t + A^2\frac{(\lambda t)^2}{2!}\right\}$	$\frac{1+A+A^2}{\lambda}$
n	$P_S(t) = e^{-\lambda t}\left\{1 + A\lambda t + A^2\frac{(\lambda t)^2}{2!} \cdots + A^n\frac{(\lambda t)^n}{n!}\right\}$	$\frac{1+A+A^2+\cdots+A^n}{\lambda}$

Table 10.5. Reliability and MTTF for system with *N* operating units, *n* standby units, and perfect changeover

No. of standbys	*System reliability (survival probability)*	*MTTF*
1	$R_S(t) = e^{-N\lambda t}(1 + N\lambda t)$	$\frac{2}{N\lambda}$
2	$R_S(t) = e^{-N\lambda t}\left\{1 + N\lambda t + \frac{(N\lambda T)^2}{2!}\right\}$	$\frac{3}{N\lambda}$
n	$R_S(t) = e^{-N\lambda t}\left\{1 + N\lambda t + \frac{(N\lambda t)^2}{2!} \cdots + \frac{(N\lambda t)^n}{n!}\right\}$	$\frac{1+n}{N\lambda}$

If the average availability of the changeover unit is assumed to be A, the system reliability $R_S(t)$ and Mean Time To Failure (MTTF) can be shown to be as given in Table 10.4.

In practice, the changeover unit might have been subject to periodic proof tests and, hence, its average availability could have been found from its inherent failure rate and the interval between tests, as described in section 10.3.

(d) Standby systems with several operating units

Consider a system in which there are N operating units and n standbys. Such a system might consist of five identical pumps three of which must work for system success. If a working pump should fail it is immediately replaced by one of the spare pumps. It is assumed that the changeover occurs without failure in negligible time and that failed units are not repaired. Such a system is shown in Fig. 10.6.

If the failure rate per unit is λ, then the average number of unit failures in time t is $N\lambda t$. The system reliabilities and Mean Times To Failure (MTTF) are shown in Table 10.5 (which may be compared with Table 10.3).

(e) Standby systems: examples

Example 10.3

Three identical pumps are arranged as in Fig. 10.7, and for each pump the mean failure rate, λ, is 5×10^{-4}/hr. Perfect changeover is assumed, and failed pumps are

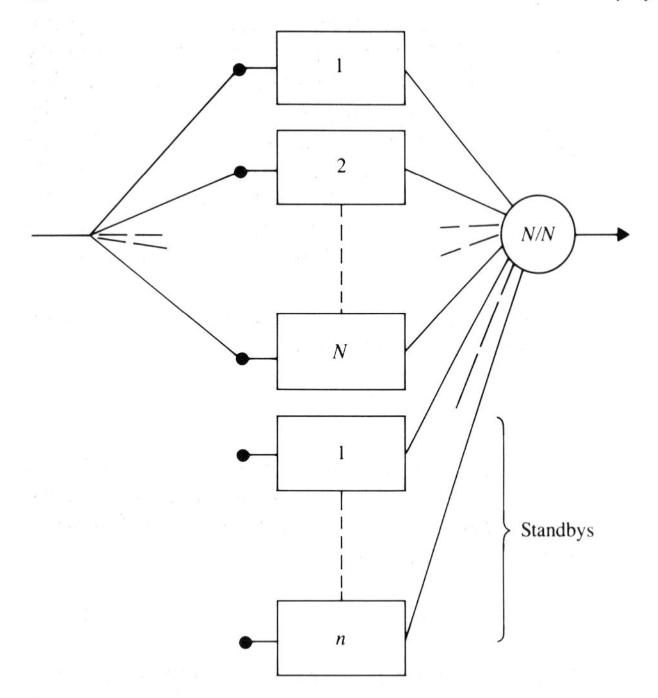

Fig. 10.6. System with *N* operating units and *n* standby units and perfect changeover

not repaired. If the system is required to run for 2×10^3 hours, what is the reliability of the system for this period and the system MTTF?

At 2×10^3 hours

$$\lambda t = 5 \times 10^{-4} \times 2 \times 10^3 = 1$$

From Table 10.3 the system reliability, $R_S(t)$, is

$$R_S(t) = e^{-\lambda t}\left\{1 + \lambda t + \frac{(\lambda t)^2}{2!}\right\}$$

$$R_S(2 \times 10^3) = e^{-1}\left\{1 + 1 + \frac{1}{2}\right\}$$

$$= 0.92 \text{ (or 92 per cent)}$$

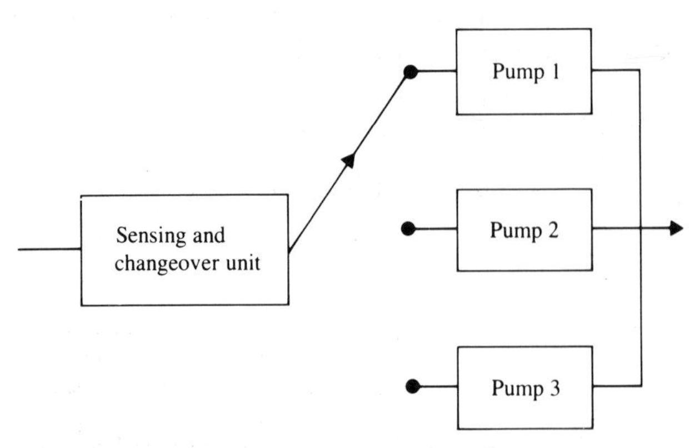

Fig. 10.7. Pump system with one operating unit and two standby units with perfect changeover: failed units not repaired

and system Mean Time To Failure is given by

$$\text{MTTF} = \frac{3}{\lambda} = \frac{3}{5 \times 10^{-4}} = 6000 \text{ hours}$$

Example 10.4

For the same system as in Example 10.3 and as shown in Fig. 10.7, what is the reliability of the system if the system is required to run for 2×10^3 hours, but the sensor-changeover unit is found to be imperfect with an availability of 0.95? What is the system MTTF?

As before, at 2×10^3 hours

$$\lambda t = 1$$

and from Table 10.4 system reliability is

$$R_S(t) = e^{-\lambda t}\left\{1 + A\lambda t + A^2\frac{(\lambda t)^2}{2!}\right\}$$

$$R_S(2 \times 10^3) = e^{-1}\left\{1 + 0.95 + 0.95^2 \times \frac{1}{2}\right\}$$

$$= 0.88 \text{ (or 88 per cent)}$$

and system MTTF is given by

$$\text{MTTF} = \frac{1 + A + A^2}{\lambda} = \frac{1 + 0.95 + 0.95^2}{5 \times 10^{-4}} = 5700 \text{ hours}$$

Example 10.5

A system consists of four identical operating pumps, all of which must be working for system success. There are two standby pumps (see Fig. 10.8), each of which can

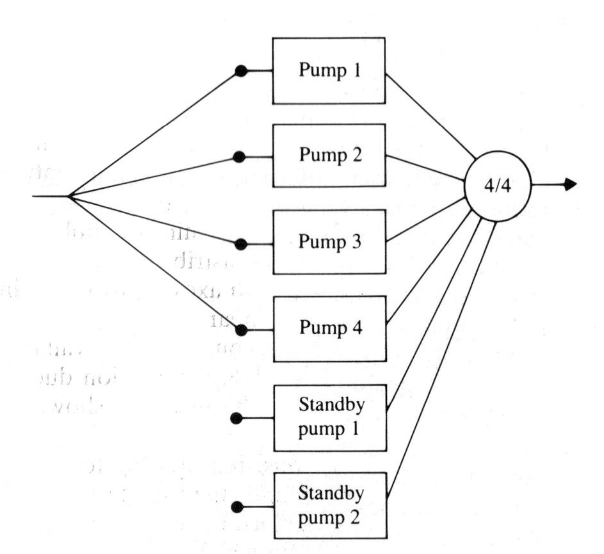

Fig. 10.8. Pump system with four operating units and two standby units with perfect changeover: failed units not repaired

be substituted immediately should an operating pump fail. For each pump, mean failure rate $\lambda = 2 \times 10^{-4}$/hr. If the system is required to run for 10^3 hours, what is the system reliability for this period?

At 10^3 hours

$$\lambda t = 2 \times 10^{-4} \times 10^3 = 0.2$$

From Table 10.5 system reliability, $R_S(t)$, is given by

$$R_S(t) = e^{-N\lambda t}\left\{1 + N\lambda t + \frac{(N\lambda t)^2}{2!}\right\}$$

$$R_S(10^3) = e^{-4\times 0.2}\left\{1 + 4 \times 0.2 + \frac{(4 \times 0.2)^2}{2!}\right\}$$

$$= 0.95 \text{ (or 95 per cent)}$$

and system MTTF is given by

$$\text{MTTF} = \frac{1+n}{N\lambda} = \frac{1+2}{4\lambda} = \frac{3}{4 \times 2 \times 10^{-4}} = 3750 \text{ hours}$$

10.3 Fractional dead time (FDT)

The systems analysed so far have each had at least one component running, so that a failure becomes immediately obvious. Many systems, such as protective interlocks, are required to remain dormant, but to respond in the event of an occasional demand. As an example, a drain valve and pump may be required to operate on receipt of a signal from a level switch when the level in a tank exceeds the normal operating range. Such systems often are provided for plant safety, so it is important to be able to estimate their likelihood of failing to operate.

Protective system failures may be grouped into two main modes, 'fail-safe' and 'fail-to-danger'.

(a) A fail-safe failure is one which is revealed immediately, but may result in spurious operation of the system. For example, if a valve, whose duty is to shut off feed flow in a process on receipt of an alarm signal, shuts if its control air supply fails, it causes a spurious, but safe, trip of the process plant and immediate indication of its failure. The system reliability with respect to fail-safe failures may be analysed using the techniques already covered.

(b) Fail-to-danger will not be revealed immediately, possibly not until the protection system fails to operate on demand. For example, a relief valve which failed to operate at the required set pressure would reveal itself only when the system became over-pressurized, possibly with disastrous consequences. To limit the possible time that the protective system could be out of action (in a failed-to-danger state) it should be proof tested at regular intervals very much less than the MTTF of the system. If a fail-to-danger is found during a proof test and repaired, no harm is done, so the primary purpose of proof-testing is to reduce the probability of a fail-to-danger only being discovered when the protective system calls unsuccessfully for a shutdown. Unfortunately, it may be possible for the proof test operator to set the protective system in a 'failed-to-danger' state when resetting the system after a test, by not following the correct test procedure exactly.

For a single item, if the test interval is T and the mean fail-to-danger rate is λ_d, such that

$$T \ll \frac{1}{\lambda_d} \quad (\text{or } \lambda_d T \ll 1)$$

it can be shown that the average fraction of time spent in the failed state, which is usually termed the *probability of failure on demand* or the *fractional dead time* (FDT), is

$$\text{FDT} \simeq \frac{\lambda_d T}{2} \tag{10.6}$$

This is, in effect, a unit unavailability, so for a series system of n units which could fail-to-danger, the system FDT is (from equation 9.10)

$$(\text{FDT})_S \simeq (\text{FDT})_1 + (\text{FDT})_2 + \cdots + (\text{FDT})_n$$

$$\simeq \frac{\lambda_1 T}{2} + \frac{\lambda_2 T}{2} + \cdots + \frac{\lambda_n T}{2} \tag{10.7}$$

Example 10.6

A system has two protective units in series, as shown in Fig. 10.9, the units having unrevealed failure rates of $\lambda_1 = 10^{-5}$/hr and $\lambda_2 = 5 \times 10^{-5}$/hr, respectively. The time interval, T, for proof tests is 1000 hours. Calculate the system fractional dead time (FDT).

For unit 1

$$\lambda_1 T = 10^{-5} \times 1000 = 0.01$$

For unit 2

$$\lambda_2 T = 5 \times 10^{-5} \times 1000 = 0.05$$

As both λTs are very much less than unity (that is, $\lambda T \ll 1$) then, from equation (10.7)

$$(\text{FDT})_S = \frac{0.01}{2} + \frac{0.05}{2} = 0.03 \text{ (or 3 per cent)}$$

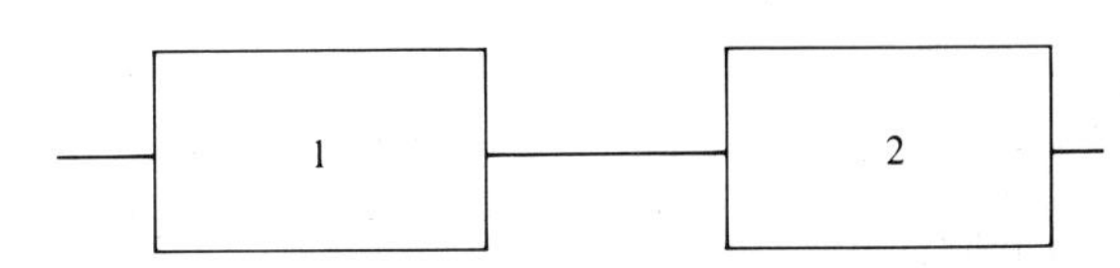

Fig. 10.9. Series system with two protective units

10.4 Fractional dead time for systems with redundancy

Often a protection system comprises several channels to reduce the probability of failure on demand. If an interlock can still operate protectively when only one out of n triggering channels remains operative, then the system FDT may be substantially reduced. However, because there are more installed devices, the incidence of spurious (fail-safe) trips increases. Consequently, 'voting systems' are usually preferred, of which the most commonly found operates such an interlock only if at least two out of the three triggering channels installed demand it. Such systems have somewhat larger FDTs than the equivalent 1-out-of-n channel systems, but have reduced spurious trip rates so that the overall failure rate of the systems is improved. In addition, for many systems, one channel at a time can be tested on-line without causing spurious trips. For a protection system of n identical channels arranged in parallel, each subject to 'unrevealed' failure at a rate, λ_d, which is proof tested at intervals of duration T such that $\lambda_d T \ll 1$, the system Fractional Dead Times $(\text{FDT})_S$ are as follows.

(a) *1-out-of-*n (system operates if at least one channel triggered). It can be shown, via an application of equation (9.18), that for the system, the FDT is given by the expression

$$(\text{FDT})_S = \frac{(\lambda_d T)^n}{n+1} \tag{10.8}$$

(b) r-*out-of-*n (system operates only if at least r channels triggered). Using equation (10.2) it can be shown that, for the system

$$(\text{FDT})_S = {}^nC_m \times \frac{(\lambda_d T)^m}{(m+1)} \tag{10.9}$$

where

$$m = n - r + 1$$

Example 10.7

A safety system (see Fig. 10.10) consists of three identical level switches monitoring the level of liquid in a holding tank. A drain valve and pump are activated if at least one switch indicates danger level. What is the FDT of this danger-level monitoring system if, for each switch, $\lambda_d = 0.5$/year and the system test interval T is 3 months (=0.25 year).

From equation (10.8)

$$(\text{FDT})_S = \frac{(\lambda_d T)^n}{n+1} = \frac{(0.5 \times 0.25)^3}{3+1} = 4.9 \times 10^{-4}$$

Example 10.8

For the same equipment as in example 10.7, a 'voting system' is installed so that drainage is not actuated unless at least two switches indicate danger-level. What is the system FDT for this 2-out-of-3 protective system?

From equation (10.9)

$$(\text{FDT})_S = {}^nC_m \times \frac{(\lambda_d T)^m}{(m+1)}$$

where

$$m = n - r + 1$$

therefore

$$m = 3 - 2 + 1 = 2$$

and

$$(\text{FDT})_S = {}^3C_2 \times \frac{(0.5 \times 0.25)^2}{2+1} = (0.5 \times 0.25)^2 = 0.0156$$

Examples 10.7 and 10.8 show that the FDT for a 1-out-of-3 system can be considerably smaller than the FDT for an equivalent 2-out-of-3 system. Conversely, the 2-out-of-3 system has a lower spurious failure rate than the 1-out-of-3 system because a system spurious failure cannot occur until two channels fail-safe.

A comparison of the fail-safe rates, fail-to-danger rates and FDTs of several different systems is shown in Table 10.6. The expressions are approximate, but hold good if the test interval is very much less than the MTTFs. The numerical values, which are illustrative only and for comparison with the previous example, were calculated assuming:

fail-safe rate per channel	$\lambda_s = 0.5$/year
fail-to-danger rate per channel	$\lambda_d = 0.5$/year
test interval	$T = 3$ months (0.25 year)

10.5 Optimization of fractional dead time (FDT)

(a) Staggered testing

The FDT of redundant (or voting) systems may be reduced if, instead of testing all the units (or channels) together at intervals of time T, the tests of individual units (or channels) are staggered. For a system with n units, the test interval for any unit remains T, but tests are carried out in succession at intervals of T/n. For various 'm out of n' protective systems, Table 10.7 gives the reduction in FDT achieved by such staggered testing, where

$$\text{FDT improvement factor} = \frac{\text{FDT (simultaneous testing)}}{\text{FDT (staggered testing)}}$$

(b) Frequency of testing

Thus far, the FDT calculations have assumed that testing time is negligible compared with the test interval, and that, therefore, the FDT continues to decrease as the testing interval is reduced. In practice, if system testing is carried out very frequently, the FDT increases because

Table 10.6. Comparison of fail-safe rates, fail-to-danger rates, and FDT for several specified protective systems

System	*Fail-safe rate (λ_s) system* Expn.	*Value (rate/yr)*	*Average interval between fail-safe (yr)*	*Fail-to-danger rate (λ_d) system* Expn.	*Value (rate/yr)*	*Average interval between fail-to-danger (yr)*	*FDT* Expn.	*Value*
1001	λ_s	0.5	2.0	λ_d	0.5	2.0	$\frac{\lambda_d T}{2}$	0.063
1002	$2\lambda_s$	1.0	1.0	$\lambda_d^2 T$	0.063	16.0	$\frac{\lambda_d^2 T^2}{3}$	0.0052
2002	$\lambda_s^2 T$	0.063	16.0	$2\lambda_d$	1.0	1.0	$\lambda_d T$	0.125
1003	$3\lambda_s$	1.5	0.67	$\lambda_d^3 T^2$	0.0078	128.0	$\frac{\lambda_d^3 T^3}{4}$	0.0005
2003	$3\lambda_s^2 T$	0.1875	5.33	$3\lambda_d^2 T$	0.1875	5.33	$\lambda_d^2 T^2$	0.0156

The system notation: 1001 means 1-out-of-1 system; 1002 means 1-out-of-2 system, etc.

the time that the system is inoperative because of testing and repair becomes comparable with the time that it is inoperative because of faults. An optimum test interval can be calculated for each system at which the FDT is a minimum. In addition, too frequent testing can increase testing costs and increase the opportunities for human errors to occur.

Example 10.9

For the 2-out-of-3 safety system shown in Fig. 10.10, the test time for each level switch is 2 hours and $\lambda_d = 0.5$/year. What is the minimum FDT and the optimum test interval?

From Table 10.6, for a 2-out-of-3 system,

$$FDT_{fault} = \lambda_d^2 T^2$$

where T = test interval in years.

Also

$$FDT_{test} = \frac{t}{T \times 8760} \quad \text{(for each switch)}$$

where t = total hours/test.

So

$$FDT_{total} = (0.5)^2 T^2 + \frac{3 \times 2}{8760 \times T} \quad \text{(for 3 switches)}$$

Table 10.7. FDT improvement factor for *m* out of *n* protective systems

m out of *n*	1	2	3	4
1	1	1.6	3.0	6.12
2		1	1.5	2.67
3			1	1.45
4				1

Differentiating this expression with respect to T gives

$$\frac{d(FDT)}{dT} = 0.5T - \frac{1}{1460T^2}$$

and the value of the right hand side of the above expression will be zero when the value of T is such as to minimize the FDT; that is, for minimum FDT

$$0.5T = \frac{1}{1460T^2}$$

or

$$T^3 = \frac{1}{730}$$

The optimum test interval, T, is therefore

$$T = 0.111 \text{ years} = 40.5 \text{ days}$$

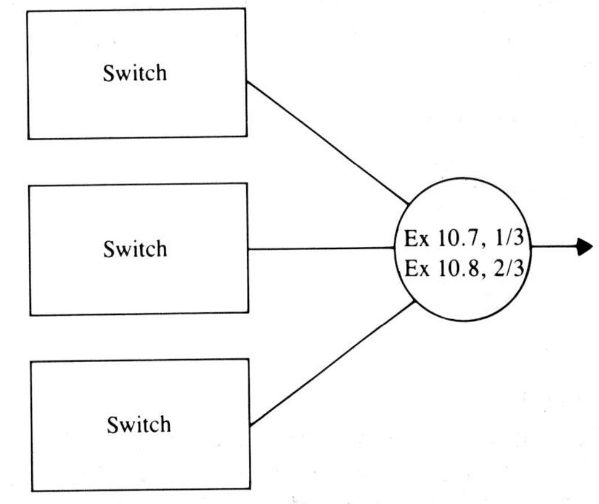

Fig. 10.10. Danger-level detection system: one out of three, or two out of three protective system

The minimum fractional dead time for the system is therefore

$$FDT_{total} = (0.5)^2 \times (0.111)^2 + \frac{6}{8760 \times 0.111}$$

$$= 0.00925$$

A fractional dead time of 0.00925 indicates that, *on average*, the expected time per year during which the system would be out of action would be $0.00925 \times 365 =$ 3.38 days/year.

PART FOUR

Techniques for process plant reliability assessment

CHAPTER 11. INTRODUCTION

11.1 Techniques covered

In Part Three the reader has been introduced to a systems approach to reliability assessment for new process plant design, through the use of Reliability Block and Logic Diagrams. The application of these basic techniques of reliability assessment to relatively simple systems has been demonstrated. Here, in Part Four, this analysis will be extended through the use of more sophisticated and flexible techniques, requiring a greater degree of judgement on the part of the assessor (Chapters 12, 13, and 14) and by the application of reliability block and logic diagrams to more complex problems (Chapter 15).

The following techniques are covered in Chapters 12, 13, 14, and 15.

(a) *Generic parts count reliability prediction*
This is a technique for predicting the reliability of equipment based upon the observed reliability of similar equipment in similar environments. Data for this observed reliability of similar equipment can be obtained either from in-house observations or from internationally available published data sources. The problems of accuracy and relevance of some of this data are fully discussed.

(b) *Fault Tree Analysis (FTA) and Failure Mode and Effect Analysis (FMEA)*
These are two complementary techniques for establishing the relationship between component or sub-assembly failures and system level failure effects. The techniques complement each other and the reliability block diagram (RBD) system modelling techniques described in Part Three.

(c) *Complex systems: some further methods of analysis*
Here the basic techniques of Part Three are extended to enable the analysis of larger or more complex systems to be carried out. Using these techniques, reliability assessments can provide predictions of availability and probability of failure for complete plant systems, even for new processes where there is no operating experience, provided that reliability data, in particular failure rates and repair times, are available for the component parts of the plant.

The techniques described in this part are intended to be used in the design phase of any project, and can be applied at any level from individual components up to complete plants or systems. When applied sensibly the techniques can help the designer to improve reliability and can reduce the amount of development testing required by identifying potential problems on the drawing board. They can also increase the designer's understanding of the factors affecting the reliability of his design and can focus attention and resources on critical areas.

Although originally developed in the aerospace, defence, and electronic industries, the techniques are equally applicable to all areas of mechanical and process engineering. Practical examples of the application of the techniques in these areas are included. Whatever the application and the technique used, the objectives of reliability design assessment remain the same.

To provide assurance that a proposed design is capable of achieving its reliability requirements.

To identify features of a design which may degrade its reliability in service.

11.2 The design process

The ideal design process is a logical progression from a problem to a solution in five distinct stages.

(a) Requirement – A simple statement of the problem to be solved.

(b) Specification – A refinement of the requirement, listing all additional constraints (for example, size, performance, environment) which will influence the choice of solution.

(c) Basic design – The identification of the best overall solution to the problem.

(d) Detail design – The refinement of the solution to the point at which it can be manufactured.

(e) Development – Corrections and improvements to the detail design to reflect the results of tests and initial in-service experience.

In any complex project it is likely that several different reliability design assessment techniques will be required at various stages of this process. Figure 11.1 shows a typical range of reliability design assessment activities during a major project. Each technique will require a commitment of manpower and other resources, and must be correctly phased with respect to other project activities. Therefore it is often necessary to prepare a reliability programme plan at the outset of a project, defining each technique to be applied, its timing, and the staff responsible for carrying out the task.

11.3 Assessment methodology

Notwithstanding the particular techniques which may be used in an assessment of a process plant, for complex systems a consistent, logical approach is essential. The plant specifications, engineering drawings, and related documentation alone can produce a pile of paper several feet in height, so that even isolating the basic information necessary to carry out an assessment can present quite a

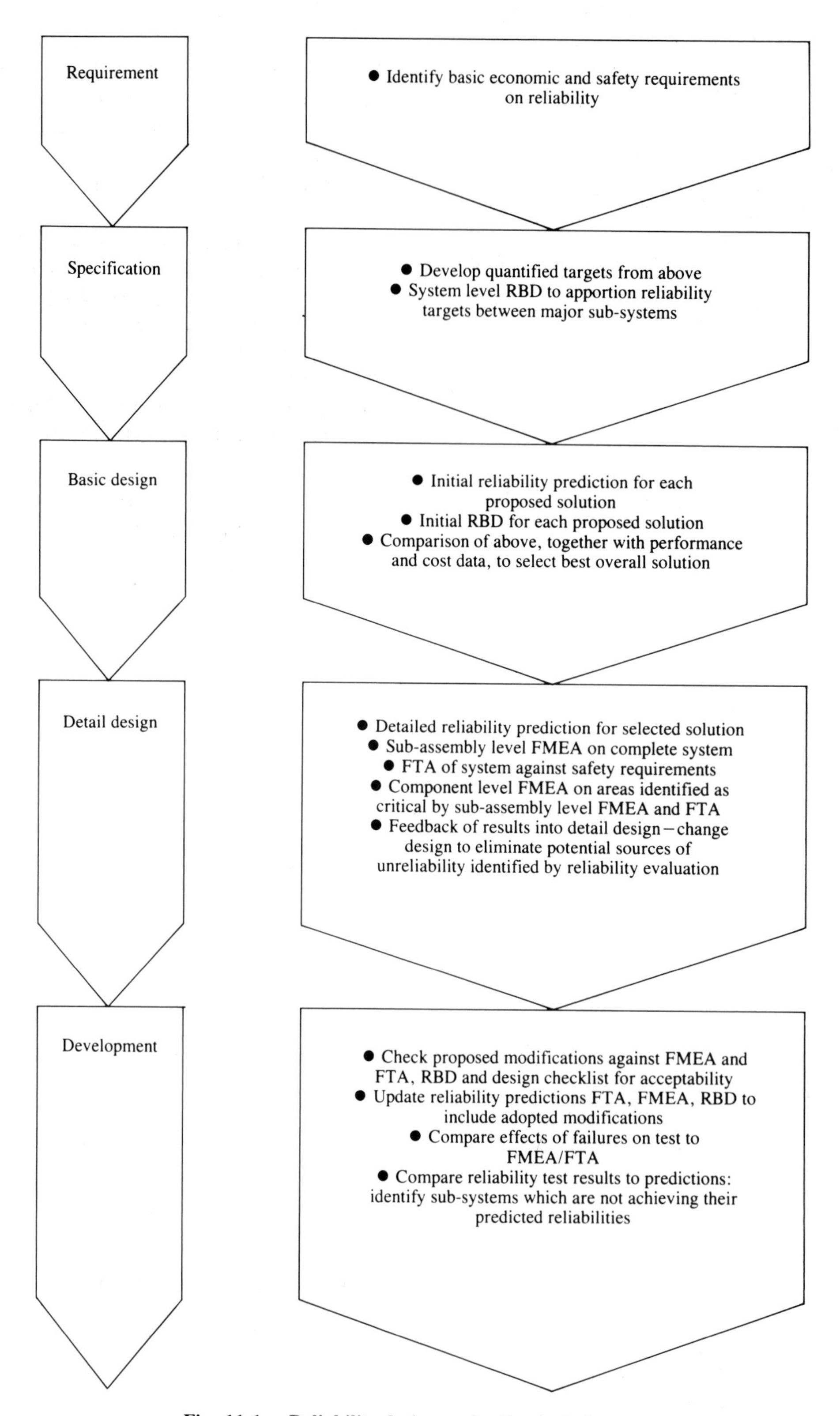

Fig. 11.1. Reliability design evaluation techniques

problem. For process plant the most suitable technique appears to be to concentrate initially on the process and instrumentation (P and I) diagrams. With these, and the plant specifications and process details as an information base, the following procedure is adopted.

(1) Define:
   (a) the assessment objectives;
   (b) the plant boundaries;
   (c) the inputs and outputs across the boundaries;
   (d) the system operating states and failure criteria;
   (e) the assumptions made in the availability model.

(2) Write out an appraisal of the plant's engineering and process characteristics in the form of a plant description.

(3) Develop a simplified flowsheet and block diagrams of the plant and sub-systems to a level at which reliability data are available.

(4) Produce failure logic networks from the block diagrams from which the overall availability model can be derived.

(5) Assess the reliability data against the defined failure states and criteria.

(6) Apply these optimized data into the availability model equations.

(7) Rank the sub-systems and components in order of their availability.

(8) Carry out a sensitivity analysis for the low-availability sub-systems and components.

(9) Review the final availability estimates in the light of published information and the experience of plant engineers, operators, and equipment specialists.

(10) Write a final report high-lighting sensitive areas and the sub-systems and components requiring further study.

CHAPTER 12. GENERIC PARTS COUNT RELIABILITY PREDICTION

12.1 Introduction

Generic parts count reliability prediction is the most widely used reliability prediction technique for mechanical and process equipment. The technique is based upon the principle that the reliability of any equipment depends upon the number of parts comprising the equipment and the reliability of each part. It is assumed for this method that the system is a series system. It is also assumed that each part has a (constant) failure rate which is the same as that observed in the past for similar parts in similar applications. These failure rates may be obtained from standard (published) databases or from in-house records. The failure rate of the equipment is then the total of these individual failure rates.

Clearly, the prediction is only as good as the data on which it is based. Many mechanical parts are unique designs for a particular application and there may be no previous reliability experience to draw on. In this case an assumed failure rate may have to be used based on judgement and experience of similar components or equipment.

Mechanical part reliability is often very sensitive to small changes on loading, duty cycle, environment, maintenance policy, and quality. Standard databases often fail to describe the source of the data in respect of these factors adequately, and, therefore, different databases often list widely differing failure rates under the same generic part description. Anyone performing a prediction or using the results must be aware of these difficulties and should treat the results accordingly. In these circumstances the relevance and importance of carrying out a sensitivity analysis on the predicted results cannot be overstressed.

12.2 Sources of generic part data

In general the most accurate reliability predictions are those based on actual field experience with similar equipment in the specific application under consideration. Many designs are derivative improvements of existing items, with the same generic component types under similar environments, maintenance policies, and so on. Therefore, the designer's first step should be to approach his customer or component supplier, or to look within his own organization for the necessary information on total usage, unscheduled removals, equipment failures, and so on. If this information is not being recorded, then the organization concerned is throwing away data which would cost many thousands of pounds to generate through specific reliability testing.

Further information on the collection and processing of data is given in Part Five, and the analysis of this in-service data was fully covered in Part Two.

In the absence of specific data from previous applications there are several published sources containing generic failure rate data for mechanical components. Some of the more commonly used sources of data are:

(a) *OREDA: Offshore Reliability Data Handbook* (**14**)
A very comprehensive and thorough presentation of reliability data for offshore systems compiled from failure and repair records already existing in company files and records. For each item, the handbook describes failure modes, failure rates for each mode with associated 90 per cent confidence interval, the number of failures on which the data is based, detailed item and application descriptions, and environmental and operational conditions.

(b) *Non-Electronic Parts Reliability Data* (**15**)
Published by the Reliability Analysis Center of the US Department of Defense, this document covers a wide range of mechanical, electro-mechanical, and hydraulic components. For each item the document presents failure rates and associated confidence limits for a number of different environments. All failure rates are based on actual field data, and in each case the source of the data (military or commercial) is indicated. Some of the generic type descriptions are very general, and in some cases there are very wide variations between failure rates quoted for similar environments. The document also contains useful data on failure modes and mechanisms for non-electronic components. It is also available in the form of a computer database program suitable for use on an IBM PC or other compatible computer.

(c) *Defence Standard 00–41, Part 3* (**16**)
Published by the Ministry of Defence, this document contains failure rates for mechanical and electro-mechanical components. The failure rates are presented as base failure rates for the 'Ground Fixed' environment, with modifying factors for other environments (for example, 'Ground Mobile', 'Ship Protected'). The generic part type descriptions are again sometimes rather general (for example, 'Heat Exchanger') and no indication of confidence limits is given. Further, the equipment covered is manufactured to Ministry of Defence Specifications which are normally more rigorous than industrial specifications.

(d) *Reliability and Maintainability in Perspective* (**17**)
This is a reliability textbook which includes tables of failure rates for a wide range of components including some mechanical items. The data has been collected from over 20 different databanks and by combining the range of values observed with the author's own experience. The presentation of information is notable in that ranges of values are given for some components, reflecting the wide variation in failure rates given for the same generic type description by different databanks.

(e) *Reliability Technology* (**18**)
This textbook includes an appendix which presents a list of electronic and mechanical component failure rates. The generic part type descriptions are again non-specific

Table 12.1. Failure rates for ball bearings from different sources

Data source	*Description/application*	*Failure rate* (FPMH)*
12.2(a)	No data	—
(b)	Bearing, spherical, ground mobile environment (60 per cent confidence interval 0.169, 0.252 FPMH)	0.206
(c)	Bearing, ball, heavy duty, ground fixed environment	4.5
(d)	Bearing, ball, heavy, ground fixed environment	2–20
(e)	Ball bearing, heavy duty, ground fixed environment (100 per cent of rating; component temperature 20°C)	10.0

* FPMH = Failures per million hours per single device.

(for example, control valves, pressure switches) and the appendix gives no indication of the size of the sample on which each failure rate is based. Some limited information on different failure modes is given. General factors to modify failure rates for environment, stress, and component temperature are given.

To illustrate the results which may be obtained from different data sources, the failure rates shown in Table 12.1 have been obtained from the above sources for a simple ball bearing.

Faced with such a range of values, the designer must apply his engineering judgement in choosing the value most appropriate to his application. This involves a consideration of size, complexity, application, and so on.

12.3 Fee-paying data sources

In addition to the published data sources referred to in section 12.2, there are a number of databanks of reliability accessible on a fee-paying or membership basis. These include the following

(a) *Systems Reliability Service (SRS)*
This is currently the most widely used reliability databank, comprising some 10 000 data sets. The databank contains failure rate information and a limited amount of failure mode information, largely obtained from the nuclear industry. Access to the databank is on an Associate Membership basis, whereby Associate Members are expected to contribute data to the databank as well as extract data for their own studies.

(b) *Hazard and Reliability Information Services (HARIS)*
This is a new service which commenced in January 1987, and provides data on a fee-paying basis. It comprises three previously separate databanks of reliability, maintainability, and accident data, together with an abstracts database. The data held includes 1400 data sheets containing failure rate data for generic equipment types, of which 30–40 per cent also include failure mode data. This databank also includes data from the offshore and process industries (1900 data sheets) and a certain amount of maintainability data (350 data sheets).

An example of the use of generic parts data is given below.

Example 12.1

The sub-system shown in Fig. 12.1 is designed to cool a flow of hydrocarbon gas. The sub-system is mounted on an offshore platform, and uses a gas/seawater heat exchanger. If cooling is not required the heat exchanger

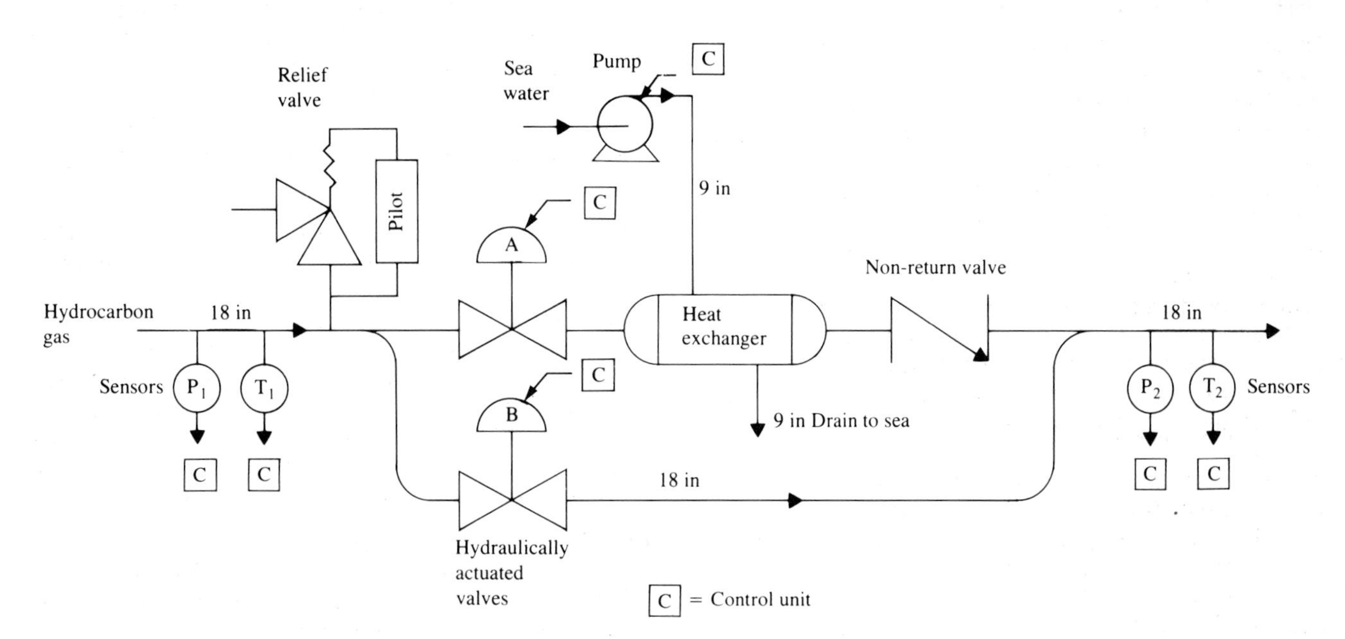

Fig. 12.1. Cooling sub-system

Table 12.2. Failure rates for the cooling subsystem shown in Fig. 12.1

Item	*Mean failure rate (FPMH*)*	*No. off*	*Total failure rate (FPMH*)*
Pressure sensor (analog electric)	42	2	84
Temperature sensor (analog electric)	12	2	24
Safety relief valve (pilot operated, hydrocarbon gas)	130	1	130
Shut-off valve (hydraulically operated, hydrocarbon gas, globe/gate/ball <24 in)	95	2	190
Heat exchanger (tube/shell type, gas/water)	99	1	99
Check valve (<24 in)	1.8	1	1.8
Pump (single stage electric motor driven, <500 kW, centrifugal, water)	260	1	260
Control unit (Prediction from Mil. Hdbk. 217D)	25	1	25
Total (FPMH):			814

* FPMH = Failures per million hours per single device.

can be by-passed, the flow being diverted by means of hydraulically actuated valves. If the gas pressure exceeds a certain pre-set value a pilot operated safety valve will open automatically, relieving the pressure. Gas pressure and temperature are monitored at the sub-system inlet and outlet. Calculate the probability of the sub-system surviving one month of continuous operation without failure.

The failure rates for the items comprising the cooling sub-system, shown in Table 12.2, were obtained from published data in the OREDA Handbook. From Table 12.2 it is seen that the total predicted failure rate for the sub-system is 814 failures per million hours (or 814×10^{-6} failures/hour). Therefore, from equation (9.11), the probability that the cooling subsystem will survive for one month of continuous operation, that is, its reliability (assuming 30 days/month), is given by

$$R(T) = e^{-\lambda t}$$
$$= e^{-814\times 10^{-6}\times 30\times 24}$$
$$= 0.577 \quad \text{(or 57.7 per cent)}$$

The piechart in Fig. 12.2 shows the relative contributions to total failure rate of different parts of the cooling sub-system.

In the case of Example 12.1 the database on which the prediction is based is well-documented and is appropriate to the application. It is instructive to note that the predicted failure rates for the same sub-system obtained from other published data sources vary from 200 to 1000 failures per million hours, from environments and component generic type descriptions that are apparently applicable. This highlights the need to base predictions on fully documented and appropriate data such as that offered by the OREDA Handbook or by the subscription databases. This aspect is considered in greater detail in Part Five.

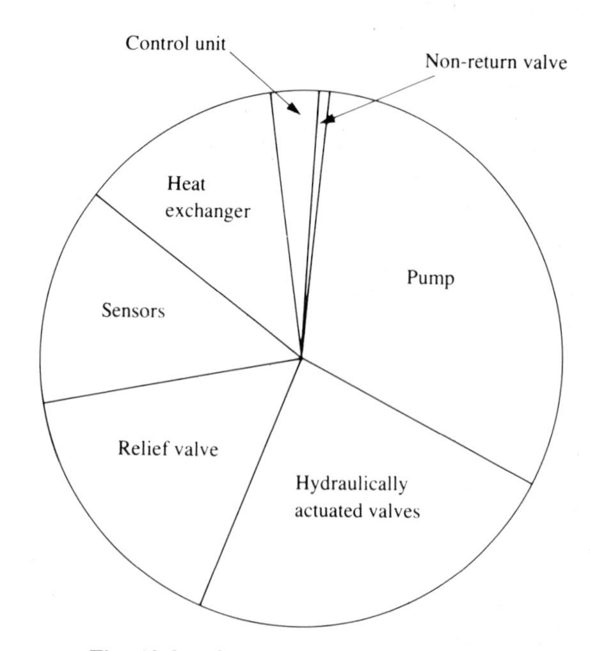

Fig. 12.2. Contributions to total failure rate

13.1 Introduction

Any process plant system will normally have a number of different 'system failure modes' (for example, leakage, loss of output, excessive output, contamination of product, and so on). Fault Tree Analysis (FTA) is a method of identifying all possible causes (that is, component failures or other events, acting alone or in combination) of a particular system failure mode, and provides a basis from which to calculate the probability of occurrence for each system failure mode. Since it can deal with combinations of component failures, it is particularly suitable for the analysis of systems containing redundancy (that is, systems containing installed spare items of equipment or sub-systems).

A fault tree shows, in graphical form, the logical relationship between a particular system failure mode (known as the TOP event) and the basic failure causes (known as the PRIME events), using 'AND' or 'OR' gate symbols. An 'AND' gate denotes that *all* the events below the gate must occur for the event above the gate to occur. An 'OR' gate denotes that *any* event below the gate will, occurring alone, cause the event above the gate to occur. The simple pumping sub-system shown in Fig. 13.1 will serve to illustrate the main principles of FTA.

Suppose that there is a need to identify causes of the system failure mode (TOP event), 'Total loss of output'. This can be caused either by blockage of the filter, failure of both pumps, or a major leakage from the interconnecting pipework. These events can be represented on a fault tree as shown in Fig. 13.2.

Circles are used to represent basic failure causes (PRIME events). The logic can be continued down to the next level, say, for 'pump failure'. This can be caused by 'failure of the electrical supply' OR 'failure of pump A' AND 'failure of pump B'. The extended fault tree is shown in Fig. 13.3.

The analysis can be continued in this way, step by step, down to whatever level of detail is required.

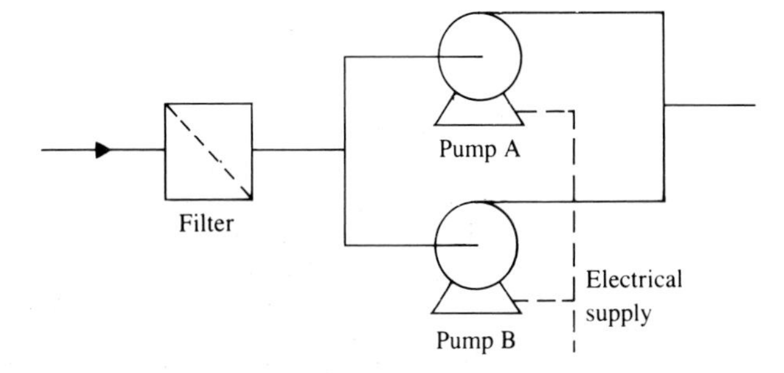

Fig. 13.1. Simple pump sub-system

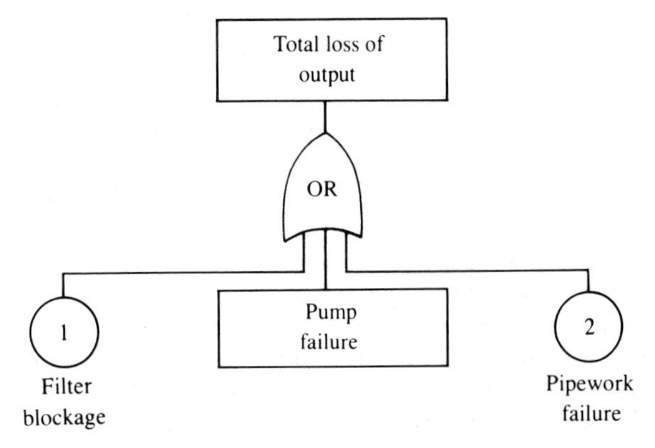

Fig. 13.2. Initial fault tree

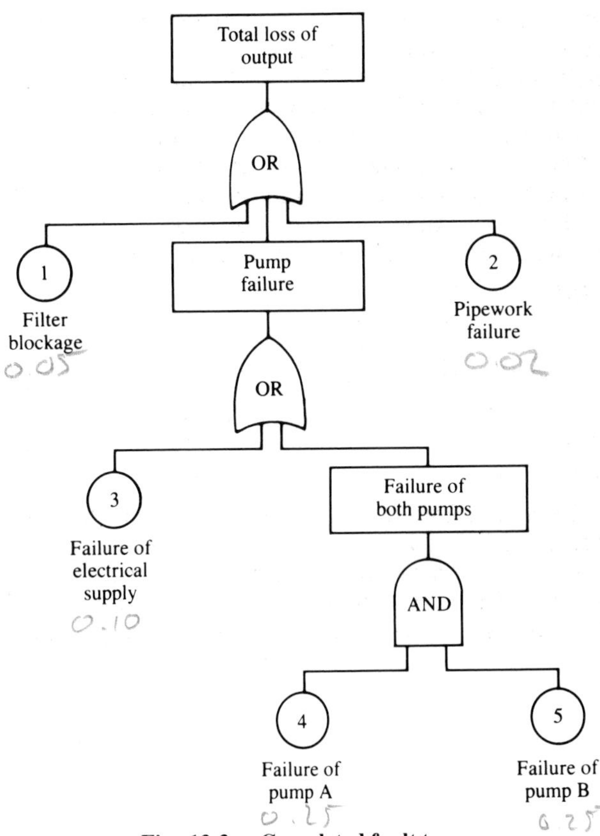

Fig. 13.3. Completed fault tree

Example 13.1

For the pumping sub-system shown in Fig. 13.1 whose fault tree is shown in Fig. 13.3 calculate the probability of total loss of output (TOP event) given that the probability of occurrence of each PRIME event is as follows:

(a) Probability of filter blockage (1) = 0.05
(b) Probability of pipework failure (2) = 0.02
(c) Probability of electrical supply failure (3) = 0.10
(d) Probability of single pump failure (4/5) = 0.25

The quantification of the fault tree proceeds gate by gate from the PRIME events to the TOP event. At the lowest level the probability of 'failure of pump A' AND 'failure of pump B' is given by equation (9.15). That is, for an active parallel system

$$F_3(t) = F_1(t) \times F_2(t)$$

Therefore, probability of failure of both pumps

$= 0.25 \times 0.25$
$= 0.0625$

At the next level, the probability of 'failure of both pumps' OR 'failure of electrical supply' is given by equation (9.3). That is, for two units in series

$$F_3(t) = 1 - [\{1 - F_1(t)\} \times \{1 - F_2(t)\}]$$

Therefore, probability of pump failure

$= 1 - \{(1 - 0.10) \times (1 - 0.0625)\}$
$= 1 - (0.9 \times 0.9375)$
$= 0.1563$

Similarly, the probability of 'filter blockage' OR 'pump failure' OR 'pipework failure', that is, probability of total loss of output

$= 1 - \{(1 - 0.05) \times (1 - 0.1563) \times (1 - 0.02)\}$
$= 1 - (0.95 \times 0.8437 \times 0.98)$
$= 0.2145$

These values, entered on the fault tree, are shown in Fig. 13.4.

Thus, from the analysis given in Part Three, to quantify an AND gate probability, the product of the individual PRIME event probabilities of occurrence is taken. Similarly, to quantify an OR gate probability, the product of the probabilities of *non-occurrence* (that is, 1 − probability of occurrence) of the individual PRIME events is taken, and the result is then subtracted from 1.

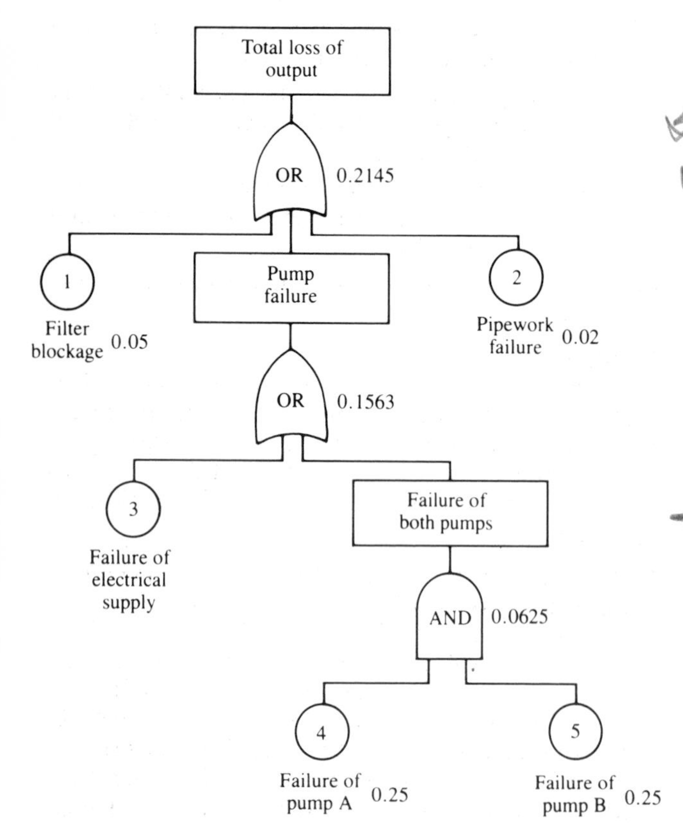

Fig. 13.4. Quantified fault tree

Computer programs are available which can be used to perform this reduction automatically in complex cases.

The procedure given here is an approximation which is quite satisfactory for simple systems. For more complex problems, where the same prime event may appear in different parts of the tree (say, loss of electric supply), then the Boolean reduction needs to be applied. This is outside the scope of this book. If the reader is interested he should refer to more advanced textbooks on Fault Tree Analysis. It is worth noting that most commercially available Fault Tree Analysis Programs automatically carry out Boolean reduction of the equations before calculating the top event probability.

The simple rules for quantification of the fault tree given above are critically dependent upon the assumption that the individual PRIME event probabilities of occurrence are completely independent. If this is not the case, the calculated probability for the TOP event may be significantly in error. For example, suppose that the two pumps of Fig. 13.1 shared a common lubricating oil system such that failure of one pump could cause contamination or loss of the lubricant, possibly leading to failure of the second pump. This is an example of a common mode (or common cause) failure and Fault Tree Analysis may be used to identify areas of a system where the possibility of common mode failures should be considered (that is, beneath AND gates). Common cause failures are considered in more detail in section 15.4, and further guidance is given in (**16**).

The technique can be applied between any levels, for example, system level down to assembly level, assembly level down to component level, and so on. However, the TOP event, and the level to which the analysis is to be taken, must always be clearly specified.

Fault Tree Analysis (FTA) is particularly suitable for use during the early stages of a project because it employs a 'top down' approach and is event-orientated. It can be used qualitatively to identify those items within a design that are likely to contribute most to a particular system failure effect and, therefore, merit closest attention. In this respect, FTA can be considered as a complementary technique to failure modes and effect analysis (FMEA – see Chapter 14). FTA can identify potentially critical items during the early design stage for deeper analysis during the detailed design stage using FMEA. A further feature of FTA is that, being event-orientated, it can include failure causes such as 'operator error'. Therefore, FTA is particularly well suited to design evaluation against safety requirements, and it is in this area that FTA finds its main use.

Fault trees provide an objective basis for analysing system design, performing trade-off studies, analysing 'common cause' (or 'common mode') failures, assessing compliance with safety requirements, and justifying design improvements or additions.

Fault Tree Analysis does have practical limitations, mainly due to the time and effort involved, particularly in a first time application. It requires very strict methodology and documentation if errors are to be avoided, and care must be taken to select the most

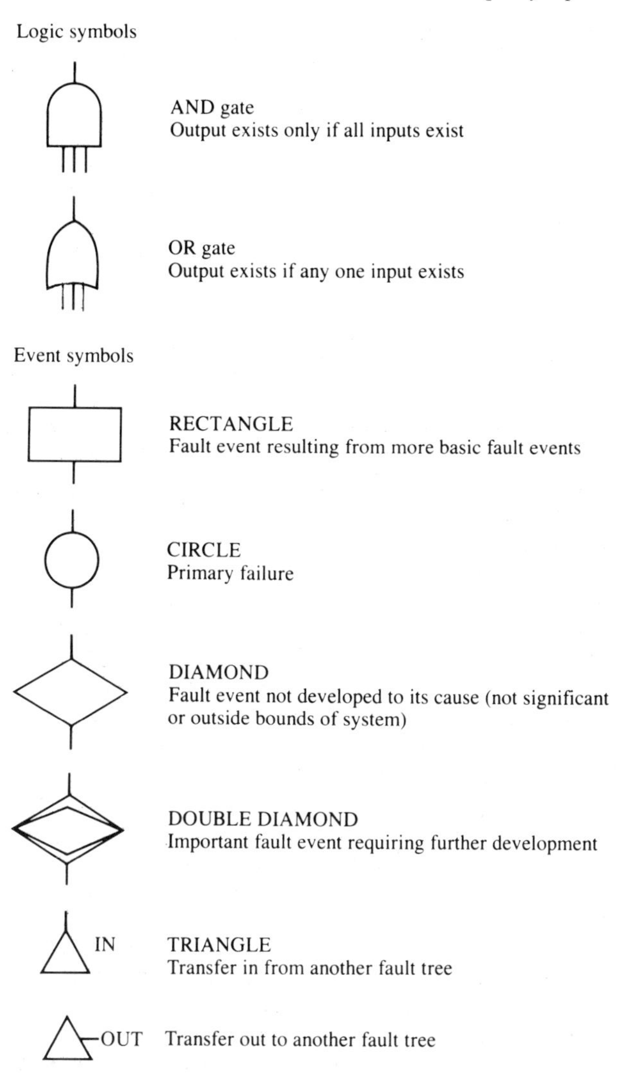

Fig. 13.5. Symbols used in Fault Tree Analysis

appropriate TOP events and levels of analysis so as to make the best use of available effort. Fault trees (or sub-trees) of more general items, or parts of a system, which may be used again in other designs should be indexed and stored for future use. A list of the more common symbols used in fault tree analysis is given in Fig. 13.5.

13.2 Method

Fault Tree Analysis involves the following main steps.

(a) Definition of the system, i.e. the items comprising the system, their functional relationships and performance requirements.

(b) Definition of the TOP event to be analysed and the bounds of the analysis.

(c) Construction of the fault tree by tracing the TOP event (failure) down to one or more causes at progressively lower levels down to the specified functional (or item) level within the design.

(d) Estimation of the probability of occurrence of each of the failure causes.

(e) Identification of any potential common cause failures affecting AND gates.

(f) Calculation of the probability of occurrence of the TOP event (failure).

It should be noted that steps (d) and (e) are desirable but not essential features of FTA. For example, for some mechanical equipment it will be difficult, or impossible, to assign quantitative values to step (d). This does not mean that FTA is valueless, as it will still identify potential areas of weakness in the design, but their relative importance will have to be assessed more on the basis of engineering judgement.

13.3 Benefits of FTA

Fault Tree Analysis benefits the design of process plant systems by:

(a) directing the analyst to discover failure deductively;

(b) indicating those parts of a system which are important with respect to the failure of interest;

(c) providing a clear and concise means of imparting reliability information to management;

(d) providing a means for qualitative or quantitative reliability analysis;

(e) allowing the analyst to concentrate on one system failure mode, or 'effect', at a time;

(f) providing the analyst and designer with a clear understanding of the reliability characteristics and features of the design;

(g) enabling the analyst to identify possible reliability problems in a design even before detailed drawings have been completed;

(h) enabling human and other non-hardware failure causes to be evaluated.

Example 13.2

Consider again the cooling sub-system shown in Fig. 12.1. Plant level safety studies have indicated that a hazard will exist if the temperature of the gas leaving this sub-system exceeds 130°C, and the probability of such an event occurring is not to exceed 1×10^{-6} per hour of operation.

In this case the TOP event is defined as 'Gas temperature at outlet exceeds 130°C' and the physical bounds of the system are defined by Fig. 12.1. The resulting FTA quantified from calculated results is shown in Fig. 13.6, sheets 1 and 2. This shows that the probability of the gas leaving the sub-system exceeding 130°C is 0.2035. In some critical cases this probability would be deemed too high and reliability improvement action would be needed to reduce the likelihood of the hazard occurring.

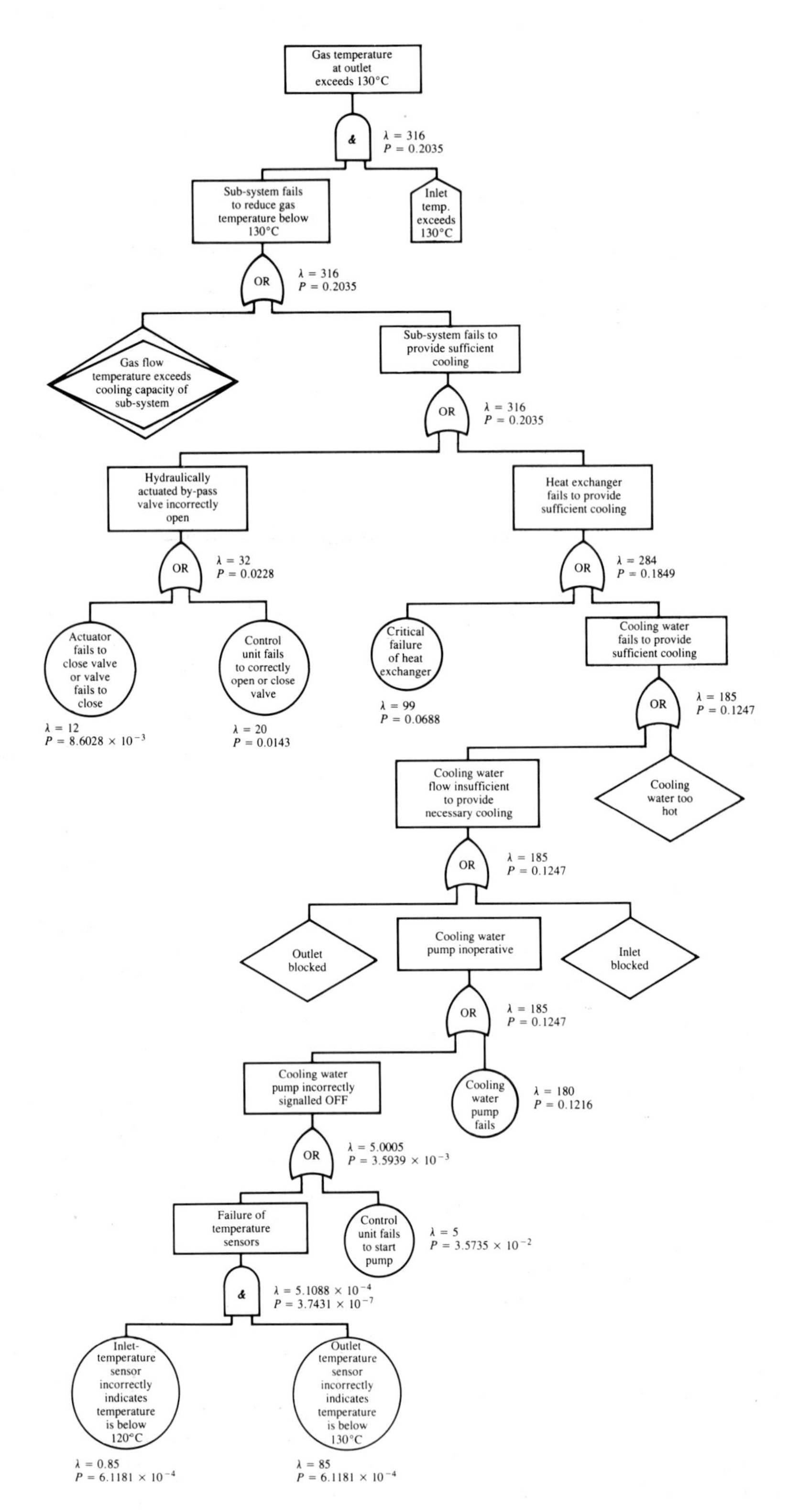

Fig. 13.6. Fault Tree Analysis of cooling sub-system

CHAPTER 14. FAILURE MODES AND EFFECT ANALYSIS (FMEA)

14.1 Introduction

Failure Modes and Effect Analysis (FMEA) is an objective method for evaluating system design by considering the various failure modes of the individual items comprising a system and analysing their effects on the reliability of that system. By tracing the effects of individual item failures up to system level, the criticality of particular items can be assessed and corrective action taken to improve the design by determining ways to eliminate or reduce the probabilities of occurrence of critical failure modes.

The analysis can be performed at any level of assembly, but the bounds of the analysis must always be clearly specified. Generally, FMEA is most effective when performed on a system which does not involve redundancy (see Chapter 9). It can then be used to identify critical areas needing additional redundant components, sub-assemblies, items of equipment, or other techniques to improve reliability.

Failure Modes and Effect Analysis is a 'bottom up' analysis as opposed to the 'top down' approach of FTA. The two techniques are complimentary, and FMEA outputs are sometimes used as inputs (PRIME events) to a higher level FTA. FMEA is usually carried out at a relatively detailed level (piece part or sub-assembly) and, thus, most commonly applied during, or after, detail design.

The use of FMEA is generally limited by the time and resources available and the capability to derive a sufficiently detailed database at the time of the analysis (for example, accurate system definition, up-to-date drawings, failure rate data, and so on). To ensure the best use of available effort, FMEA should generally be confined to items shown to be critical by earlier analyses (for example, FTA, Reliability Block Diagram (RBD) analysis, and so on) or by some other criterion (for example, safety, high cost, complexity, difficult access for maintenance, and so on).

14.2 Method

Failure modes and effect analysis involves much detailed and time-consuming work and it must be fully documented to provide a clear and well-related hierarchy of data. The procedure involves the following activities.

(a) *System definition.*
Define the system to be evaluated, the functional relationships of the items in the system and the level of analysis to be performed.

(b) *Failure modes analysis.*
Define all potential failure modes to be evaluated at the lowest level of assembly. Examples to be considered are:

premature operation;
failure to operate at a prescribed time;
intermittent operation;
failure to cease operation at a prescribed time;
loss of output or failure during operation;
degraded output.

(c) *Failure effects analysis.*
Define the effect of each failure mode on the immediate function or assembly, each higher level of assembly, and the function/mission to be performed. This could include a definition of the symptoms available to the operator.

(d) *Rectification (optional).*
Define the immediate action required by the operator to limit the effects of the failure or to restore an immediate operational capability, plus the maintenance actions required to rectify the failure (providing the basis for subsequent maintainability analyses).

(e) *Quantify failure rate (optional).*
If sufficient data exists, the failure rate or failure probability for each failure mode should be defined. The total failure rate or failure probability associated with a particular failure effect can then be quantified.

(f) *Criticality analysis (optional).*
Define a measure combining the severity of the failure effects with the probability of the failure occurring. The criticality analysis can be quantitative or qualitative. For example, the analyst could identify the failure effects by number: 1 (hazardous failure) to 5 (no effect), and could make a subjective assessment of failure probability: A (possible), B (unlikely), C (extremely unlikely). All failure modes identified as 1A would then be identified as critical failures for urgent corrective action (see (g)).

(g) *Corrective action (optional).*
Define changes to the design, operating procedures or planned testing to limit the effects and/or to reduce the probability of critical failures.

The FMEA is usually carried out on a *pro forma* with separate columns for each of the items (a) to (g) listed above. The design of the *pro forma* and the range of activities involved may well vary from system to system, to reflect the particular objectives of the analysis. It is vital that these objectives should be fully defined from the start so that the best possible return is obtained from the considerable investment of time and effort in carrying out such an analysis.

14.3 Functional FMEA

The choice of level of analysis to be performed is particularly important in FMEA of mechanical designs. The analyst must consider the resources available, the level and detail of failure rate data, and the ease and accuracy with which failure modes and effects at various levels can be defined. It is often better to produce a short FMEA which considered failure modes at sub-assembly level with clearly defined failure effects rather than an extremely lengthy and tedious piece-part level analysis.

A functional FMEA is a further alternative to a lengthy piece-part level analysis. A functional FMEA is based on the functional structure of the system rather than on its physical components. The functional approach should be used either if the items do not have a physical identity or if the system is complex. The functional FMEA is identical to the normal FMEA, except that failure modes are expressed as 'failure to perform particular sub-system functions'. The functional FMEA should consider both primary functions (that is, the functions for which the sub-system was provided) and secondary functions (that is, functions which are merely a consequence of the sub-system's presence). An example of a secondary function of a hydraulic valve would be 'to maintain hydraulic fluid cleanliness'. Failure of the sub-system to perform this function may have no effect on the primary functions of the system, but could have severe consequences for other systems using the same hydraulic supply.

Example 14.1

Consider again the cooling sub-system shown in Fig. 12.1. Suppose that an FTA had identified failures of the pilot operated relief valves as being particularly critical to safety, and that it has, therefore, been decided to carry out a detailed FMEA of this valve. Figure 14.1 shows the internal arrangement of the valve, and the resulting FMEA is shown in Table 14.1.

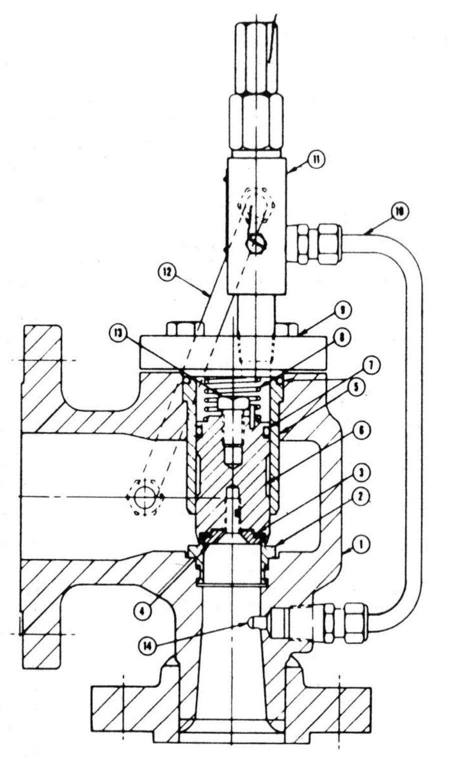

(1) Body (2) Nozzle (3) Seat (4) Seat retainer (5) Liner (6) Piston (7) Piston seal (8) Shipping spring (9) Cap (10) Supply tube (11) Pilot valve (12) Exhaust tube (13) Life adjustment screw (14) Dipper tube

Fig. 14.1. Pilot operated relief valve

Table 14.1. FMEA of pilot-operated relief valve (component level)

Failure modes and effects analysis — *System/equipment:* — *Analyst: RTP* — *Dwg no.: —* — *Date: 1.5.86* — *Issue: 1* — *Ref: Example* — *Sheet 1 of 1*

Item/ component	*Failure mode*	*Local failure effects*	*System failure effects*	*Symptoms*	*Rectification*	*Corrective action*
Seat	Scored or damaged	Relief valve leaks	Hydrocarbon gas leaks from system and pressure falls	Noise of leaking gas. Pressure reduction	Replace valve	
Piston seal	Leaking or damaged (e.g. chemical reaction)	Pressure above piston drops and relief valve opens spuriously	Hydrocarbon gas spuriously vented from system	Spurious opening of relief valve	Replace valve	Choose materials resistive to chemical attacks, etc.
Piston liner	Scored or damaged	As above	As above	As above	As above	Abrasion resistive materials for piston lining
Shipping spring	Weakened, worn or fatigued	Reduction in downwards force on piston may cause spurious, intermittent opening of relief valve	As above	As above	As above	Choose material with suitable fatigue characteristics
Supply tube	Leakage at connections	Reduction in supply pressure to pilot valve will cause increase in pressure required to activate relief valve	Hydrocarbon gas allowed to rise to a higher pressure than specified. May be critical	Gauges indicate unrelieved pressures	Replace valve	Ensure connections are clean and adequately sealed
Pilot valve	(i) Leakage	Greater pressure required to activate relief valve	As above	As above	As above	As all of those above
	(ii) Valve seized shut	Pilot valve unable to activate opening of relief valve	Hydrocarbon gas pressure allowed to rise to dangerous levels. Will be critical	Gauges indicate dangerously high pressures	As above	Careful design of pilot valve should reduce occurrence of this event to a minimum
Exhaust tube	Leakage at connections	After relief valve has been opened, unable to establish pressure equilibrium required to reclose valve	Once opened, relief valve permanently open and hydrocarbon gas continually vented from system	Relief valve fails to reclose	Replace valve	Ensure connections are clean and adequately sealed
Dipper tube	Blockage	Loss of supply pressure to pilot valve. Pilot valve unable to activate opening of relief valve	Hydrocarbon gas pressure allowed to rise to dangerous levels. Will be critical	Gauges indicate dangerously high pressure	As above	Some form of filtering may be considered to prevent blockage due to foreign objects

Table 14.2. FMEA of water chiller unit (assembly level)

WATER CHILLER UNIT *ANNEX 'A'*

Indenture level: 3
Sheet No.: 15 of 20
Mission Phase 1 – OPERATING

Study by: *R M Consultants Ltd*
Prepared by: *J. A. Butler*
Approved by: *A. O. Vanton*
Date: *5/7/85*

System Ref–Description–Function	*Failure*		*Possible causes*	*Symptom detected by*	*Effect of failure*		*Compensating provision against failure*	*Severity class*	*Remarks*
	Entry code	*Mode*			*Local*	*On next level*			
1.5.1 Vane actuator and linkage	1511	Fail open	Close direction circuit interrupted. Failed capacitor	Low chilled water temperature. Low temperature warning	Vanes stay too wide open. Hot gas bypass not available	Excess compressor throughput	Low chilled water outlet temperature protection shutdown	4	Compressor will not restart due to 'vanes closed' interlock
Position vanes in response to command. Has open and closed limit switches. Drives potentiometer & operates vanes microswitch and control transfer limit switch	1512	Linkage disconnected	Vibration effects. Incorrect assembly. Accidental damage	Low chilled water temperature. Low temperature warning	Vanes fly wide open	Excess compressor throughput	Motor thermal protection unit. Low chilled water outlet temperature protection shutdown	4	Recommend adequate locking on all threaded parts and protection against damage
	1511	Fail closed	Open direction circuit interrupted. Failed capacitor	Loss of performance. 2-hourly checks	Vanes too far closed. Hot gas bypass may stay open	Inadequate compressor throughput	2-hourly reading checks	4	Other chillers may take up load. Recommend high chilled water outlet temperature warning
	1514	Open limit switch opens too soon	Drift of setting. Switch open circuit	If severe, by study of 2-hourly readings	Vane motor fails to fully open vanes	Maximum output of compressor not available	None	4	Recommend add high chilled water temperature warning. Recommend routine check of operation
	1515	Open limit switch does not open	Drift of setting. Contacts welded	Not detected	None, because potentiometer senses position & controller stops motor	None, unless together with control or potentiometer failure	None	4	Recommend routine check of operation

Table 14.3. FMEA of water chiller unit – summary sheet

Table 10 Sheet 2 of 3
Severity category 3

Failure Modes Grouped by Severity
Compiled by J. A. Butler: Approved by A. O. F. Vanton

Date: July 1985

System component	*Failure mode*	*Possible causes*	*Entry code*	*Sev. cat.*	*Failure mode* $\lambda_m = x\lambda p$	*Eng. judgement factor* β	$\beta\lambda_m = C_m$	*Failure mode criticality Cat. 3*
Compressor	Bearing failure	Lubrication. Wear	1321	3	0.5	1.0	0.5	0.0206
Sump oil heaters	Fail to heat	Open circuit	1421	3	1.68	0.1	0.168	0.0069
Jet oil pump	Relief valve open	Spring. stuck	1431	3	0.2	0.5	0.10	0.0041
Jet oil pump	Relief valve closed	Stuck	1432	3	0.2	0.3	0.06	0.0025
Jet oil pump	Flow diverter up	Stuck	1433	3	2.0	1.0	2.0	0.0823
Jet oil pump	Flow diverter down	Stuck	1434	3	2.0	0.25	0.5	0.0206
Low speed oil pump	Low output	Blockage	1451	3	1.05	1.0	1.05	0.0432

Example 14.2

Example 14.1 was carried out at a level of detail which is not normally economic or practicable, other than in particular areas or assemblies shown to be critical by other analyses. Tables 14.2 and 14.3 show extracts from an actual process industry FMEA, in which failure modes are considered at the level of complete modules or assemblies (for example, valves, actuators, pressure vessels). Table 14.2 shows a typical FMEA *pro forma*, with a column for recording severity classification. The severity classification adopted in this case is as follows.

Category 1: Catastrophic – May cause death or system loss.

Category 2: Critical – May cause severe injury or damage.

Category 3: Major – May cause minor injury or damage.

Category 4: Minor – Unscheduled maintenance only.

Table 14.3 shows how the results of such an analysis can be grouped together by severity classification. The predicted failure rate of each system component (λ_p) is multiplied by a failure mode apportionment factor (x) to obtain a failure rate for each particular failure mode (λ_m). This is then multiplied by a factor, β, which represents an engineering judgement of the likelihood of the failure mode causing the described effect. The result is the failure mode criticality, a number which may be used to identify critical components and to set priorities for corrective design actions or increased levels of spares inventory, maintenance, and so on.

CHAPTER 15. COMPLEX SYSTEMS: SOME FURTHER METHODS OF ANALYSIS

15.1 Introduction

The concept of Reliability Block and Logic Diagrams was introduced in Part Three for the analysis of the most simple systems, consisting of a relatively few units, which could be represented by simple reliability block diagrams with straightforward series, active parallel, or inactive parallel connections. In practice, however, an operating process plant will consist of numerous units and sub-units and the resultant reliability block diagram might be a large network of complex interconnections representing complex reliability interdependencies. In the early part of this chapter, therefore, we will examine methods for reducing and/or systematising analysis of such a reliability block diagram. Furthermore, the reliability/availability of a real plant will be strongly influenced by outage times or repair rates, which are themselves functions of spares and repair-team availability, job priority rules, inspection policy, and so forth. An indication of analytical methods which can, to some extent, take account of repair related factors, will be given in the later part of the chapter.

15.2 System reduction

(a) The basic method

The analysis of a large block diagram may be carried out in successive stages. At each stage small groups of units are analysed using the methods described in Chapters 9 and 10, allowing the overall block diagram to be redrawn in a simplified and reduced form. The following example illustrates this process.

Example 15.1

The Reliability Block Diagram of a process plant system is shown in Fig. 15.1(a). The number in the bottom right hand corner of each numbered block is the assessed average availability for the unit represented by the block. The system functions satisfactorily if at least one process line through the system is operational.

The system availability is found by 'reducing' the reliability block diagram in stages, as follows.

(1) *Stage 1*

On the original reliability block diagram, Fig. 15.1(a), take each series-connected group of units, calculate its availability and assign that value to an equivalent single unit which replaces the original group.

For each of the groups (2, 3), (4, 5) and (6, 7)

$$\text{Availability} = 0.90 \times 0.80 \quad \text{(using equation (9.7))}$$
$$= 0.72$$

Similarly, for each of groups (8, 9, 10) and (11, 12, 13)

$$\text{Availability} = 0.95 \times 0.90 \times 0.80$$
$$= 0.684$$

From these results the Reliability Block Diagram can be redrawn as Fig. 15.1(b).

(2) *Stage 2*

On the reduced diagram, Fig. 15.1(b), take each parallel-connected group of units, calculate its availability and assign that value to an equivalent single unit as in Stage 1.

For group (2–7) (using equation (10.3))

$$\text{Availability} = {}^3C_2 \times (0.72)^2 \times (1 - 0.72) + {}^3C_3 \times (0.72)^3$$
$$= (3 \times 0.52 \times 0.28) + (1 \times 0.37)$$
$$= 0.81$$

For group (8–13) (using equation (9.21))

$$\text{Availability} = 1 - (1 - 0.684)^2$$
$$= 0.90$$

The system Reliability Block Diagram can now be redrawn as Fig. 15.1(c).

(3) *Stage 3*

On Fig. 15.1(c), reduce the series-connected group, as in Stage 1.

For group (1–7) equation (9.7) gives

$$\text{Availability} = 0.95 \times 0.81$$
$$= 0.77$$

The system Reliability Block Diagram is now reduced to Fig. 15.1(d).

(4) *Stage 4*

On Fig. 15.1(d), the parallel-connected group may be reduced, as in Stage 2.

For group (1–13) equation (9.21) gives

$$\text{System availability} = 1 - (1 - 0.77)(1 - 0.90)$$
$$= 0.98$$

Four stages of reduction were sufficient to solve this example. For more complicated diagrams, further similar stages may be necessary to reduce the Reliability Block Diagram to the simplest form. The reduction process should be continued until the diagram is reduced to a single equivalent unit (whose reliability or availability will be that of the system) or until no further reduction is possible using this procedure (when the conditional probability technique described in the following section may be of use).

(b) Conditional probability technique

In order to increase system flexibility, and, hence, reliability, crossover paths or cross-linkages are often provided so that some units are in more than one subsystem. The reliability block diagram cannot then be fully reduced in the manner just described. One way of dealing with this problem is to use a theorem on conditional probability, sometimes referred to as Bayes' equation. This states that:

If the occurrence of an event A depends on one of two mutually exclusive and complementary events, B_1 and B_2,

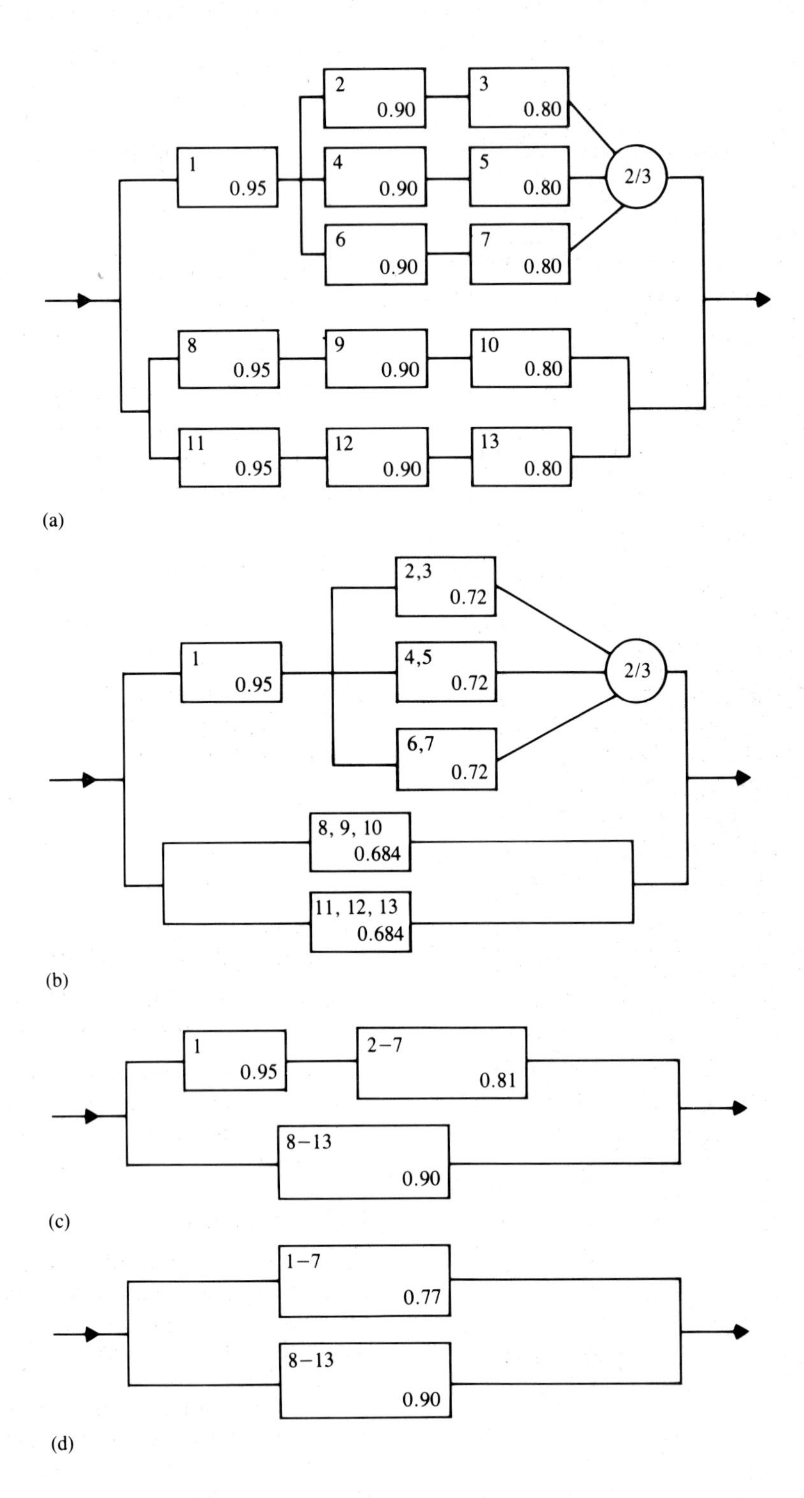

Fig. 15.1. Example of successive system reduction

then the probability $p(A)$ of A occurring is given by the expression

$$p(A) = p(A|B_1) \times p(B_1) + p(A|B_2) \times p(B_2) \qquad (15.1)$$

where

$p(A|B_1)$ = probability of A occurring given that B_1 occurs
$p(B_1)$ = probability that B_1 occurs.

(Explanatory note: Think of $p(A)$ as the probability that horse A will win the 2.30 and $p(B_1)$, $p(B_2)$ as the respective probabilities that the going will be hard or not hard).

Example 15.2

A solid product is made in batch reactors and is then removed and dried. Two driers operate in parallel and each is fed by its own conveyor. There is a third conveyor which can transfer material to either drier when required. The average unit availabilities are as shown on Fig. 15.2(a), which is the schematic diagram of the driers and conveyors. We need to calculate the average system availability at full output.

Let

A_s = Availability of system
A_1, A_2, etc. = Availabilities of unit 1, unit 2, etc.
$A_s|5$ up = Availability of system when unit 5 is up
$A_s|5$ down = Availability of system when unit 5 is down

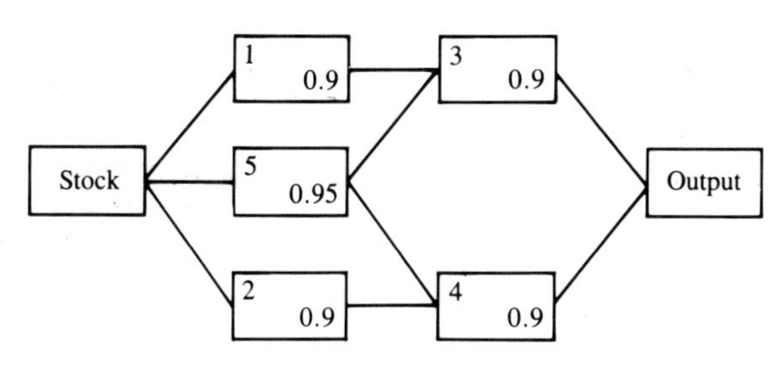

(a) Schematic diagram

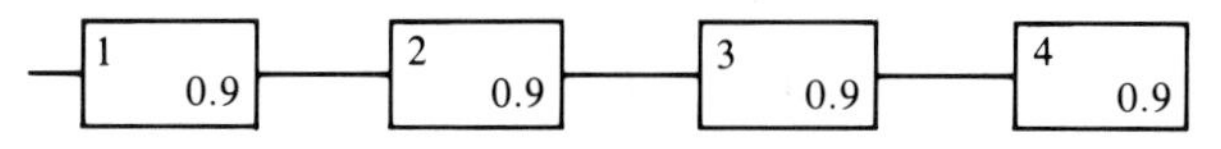

(b) Reduction if unit 5 is DOWN

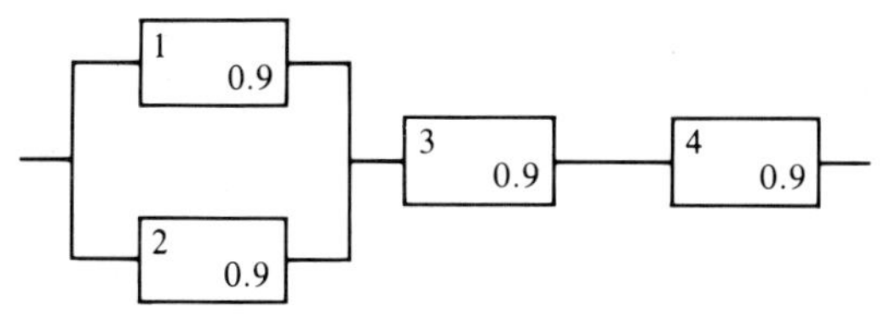

(c) Reduction if unit 5 is UP

Fig. 15.2. Example of conditional probability (Example 15.2)

Equation (15.1) can then be applied in the form

$$p\,(\text{system up}) = p\,(\text{system up}|\text{unit 5 up}) \times p\,(\text{unit 5 up}) + p\,(\text{system up}|\text{unit 5 down}) \times p\,(\text{unit 5 down})$$

or, in terms of availability

$$A_s = (A_s|5 \text{ up}) \times A_5 + (A_s|5 \text{ down}) \times (1 - A_5) \qquad (15.2)$$

The value of A_5 is known, so in order to evaluate this expression we need to evaluate ($A_s|5$ up), the system availability when unit 5 is up, and ($A_s|5$ down), its value when unit 5 is down.

The analysis is carried out in three steps as follows.

(1) *Step 1:* ($A_s|5$ down)
If unit 5 is down, units 1, 2, 3, and 4 must be up for full output. Figure 15.2(a) therefore reduces to Fig. 15.2(b), so

$$A_s|5 \text{ down} = A_1 \times A_2 \times A_3 \times A_4 = 0.9 \times 0.9 \times 0.9 \times 0.9 = 0.656 \quad \text{(from equation (9.7))}$$

(2) *Step 2:* ($A_s|5$ up)
If unit 5 is up, then, for full output, units 3 and 4 must still both be up, but only one of units 1 and 2 is essential. Figure 15.2(a) therefore reduces to Fig. 15.2(c), so

$$A_s|5 \text{ up} = \{1 - (1 - A_1)(1 - A_2)\} \times A_3 \times A_4 \quad \text{(see equation (9.21))}$$
$$= \{1 - (1 - 0.9)^2\} \times 0.9 \times 0.9 = 0.802$$

(3) *Step 3*
Substituting these values back into equation (15.2) gives

$$A_s = 0.656 \times 0.05 + 0.802 \times 0.95 = 0.795 \quad \text{(or 79.5 per cent)}$$

A large and complex Reliability Block Diagram containing numerous cross-links can be analysed by the successive application of equations such as (15.2), that is, by systematic reduction using Bayes' equation.

15.3 Truth tables

The most direct method of analysing a large Reliability Block Diagram is via a *Truth Table*. In the basic, most unsophisticated application, all the possible states in which the system represented by the diagram can exist are tabulated. For each such state the corresponding system status (for example, whether the system is operational or failed) and the probability of occurrence are also entered on the table. The total probability of the system being operational is then found by summing the probabilities of occurrence of each state identified as operational. The method may be applied to both availability and reliability (survival probability) problems.

Example 15.3

For comparison, a Truth Table is used to solve Example 15.2 which was used to illustrate the conditional probability technique in section 15.2. For the system represented by Fig. 15.2(a), the system availability is required at full output. In addition, the overall system availability is required if the system can also work at 50 per cent throughput when only one conveyor/drier path is running.

The truth table is drawn up as follows:

(1) *Unit status*
Draw up a complete set of all possible groups of states of the five items (conveyors and driers) in Fig. 15.2(a). In Table 15.1 the first 16 lines show the possible states of items 1–4 when item 5 is down, and (taken together) correspond to Step 1 of Example 15.2. The second 16 items show the same set of states for items 1–4 with item 5 up, and correspond to Step 2 of Example 15.2.

(2) *System status*
For each group of item states determine whether the system would be up or down by inspection of the system diagram, Fig. 15.2(a). The answer can depend on whether 100 per cent or 50 per cent is being considered.

(3) *State probability*
Calculate the probability of occurrence of each group of states. For example, for group AAAUU,

$$\begin{aligned}\text{Probability} &= A_1 \times A_2 \times A_3 \times (1 - A_4) \times (1 - A_5)\\ &= 0.9 \times 0.9 \times 0.9 \times (1 - 0.9) \times (1 - 0.95)\\ &= 0.0036\end{aligned}$$

In Table 15.1 probabilities are shown only for up-states. Where the group is available only for 50 per cent

Table 15.1. Truth table for Example 15.3

Unit status					*System status*			*At system output*	
1	2	3	4	5	50%	100%	*State probability*	50%	100%
A	A	A	A	U	A	A	0.9 × 0.9 × 0.9 × 0.9 × 0.05		0.0328
A	A	A	U	U	A	U	0.9 × 0.9 × 0.9 × 0.1 × 0.05	0.0036	
A	A	U	A	U	A	U	0.9 × 0.9 × 0.1 × 0.9 × 0.05	0.0036	
A	A	U	U	U	U	U			
A	U	A	A	U	A	U	0.9 × 0.1 × 0.9 × 0.9 × 0.05	0.0036	
A	U	A	U	U	A	U	0.9 × 0.1 × 0.9 × 0.1 × 0.05	0.0004	
A	U	U	A	U	U	U			
A	U	U	U	U	U	U			
U	A	A	A	U	A	U	0.1 × 0.9 × 0.9 × 0.9 × 0.05	0.0036	
U	A	A	U	U	U	U			
U	A	U	A	U	A	U	0.1 × 0.9 × 0.1 × 0.9 × 0.05	0.0004	
U	A	U	U	U	U	U			
U	U	A	A	U	U	U			
U	U	A	U	U	U	U			
U	U	U	A	U	U	U			
U	U	U	U	U	U	U			
A	A	A	A	A	A	A	0.9 × 0.9 × 0.9 × 0.9 × 0.95		0.6233
A	A	A	U	A	A	U	0.9 × 0.9 × 0.9 × 0.1 × 0.95	0.0693	
A	A	U	A	A	A	U	0.9 × 0.9 × 0.1 × 0.9 × 0.95	0.0693	
A	A	U	U	A	U	U			
A	U	A	A	A	A	A	0.9 × 0.1 × 0.9 × 0.9 × 0.95		0.0693
A	U	A	U	A	A	U	0.9 × 0.1 × 0.9 × 0.1 × 0.95	0.0077	
A	U	U	A	A	A	U	0.9 × 0.1 × 0.1 × 0.9 × 0.95	0.0077	
A	U	U	U	A	U	U			
U	A	A	A	A	A	A	0.1 × 0.9 × 0.9 × 0.9 × 0.95		0.0693
U	A	A	U	A	A	U	0.1 × 0.9 × 0.9 × 0.1 × 0.95	0.0077	
U	A	U	A	A	A	U	0.1 × 0.9 × 0.1 × 0.9 × 0.95	0.0077	
U	A	U	U	A	U	U			
U	U	A	A	A	A	U	0.1 × 0.1 × 0.9 × 0.9 × 0.95	0.0077	
U	U	A	U	A	A	U	0.1 × 0.1 × 0.9 × 0.1 × 0.95	0.0009	
U	U	U	A	A	A	U	0.1 × 0.1 × 0.1 × 0.9 × 0.95	0.0009	
U	U	U	U	A	U	U			
								0.1941	0.7946

A = Unit, or system, available (up).
U = Unit, or system, unavailable (down).

throughput, the calculated probability is entered only in the 50 per cent output column; where the group is available for both 100 per cent and 50 per cent output, the calculated probability is entered only in the 100 per cent column.

(4) *System availability*
For full output this is given by the sum of probabilities of '100 per cent output' states = 0.795 (which is, as it should be, the result previously obtained by the conditional probability technique used in Example 15.2).

If the plant continues to function at 50 per cent output when one drier has failed, the overall availability of the plant is the availability at 100 per cent output plus half of the additional availability at 50 per cent output, that is

$$A_{\text{Total}} = 0.7945 + 0.5 \times 0.1941 = 0.8916$$

Clearly, the disadvantage of the Truth Table method is that, even with the simplest of modelling assumptions, that is, with two states only (unit either operational or failed) as in the example, a system of N units will generate a table with 2^N entries, so that each extra unit doubles the length of the calculation. However, it can be computerised and can include partial failures of units if required. If unit Unavailabilities, Fractional Dead Times, or failure-probabilities are small, computing time may often be drastically reduced with little loss of precision, by ignoring those highly unlikely system states which include more than, say, three or four unit failures.

15.4 Common cause failure (CCF)

The assumption has been made so far that the failures of individual subsections of parallel redundancy systems are independent of each other, that is that two or more subsections do not fail simultaneously from precisely the same cause (except purely by chance). However, most systems have the potential of having more than one failure due to a common cause in a significantly short period, and this limits the improvement in reliability which can be achieved by the provision of parallel redundancy systems. Common Cause Failures (CCF) tend to arise from errors made during design, manufacture, or operation and maintenance of a plant, or from unforeseen environmental effects. Examples can be as diverse as design or manufacturing defects in a batch of pumps used on the same plant, identical maintenance or calibration errors made by the same mechanic, and flooding caused by burst pipework or exceptional rainfall. Because these failure modes may appear to be outside the system being assessed they can easily be overlooked, leading to too-optimistic an assessment. Where they can be identified, it may be possible to eliminate them (during the design) or model them in the analysis (if the plant is already in service). For example, if loss of power supplies can cause both a running and a standby pump to fail, this common cause may be designed out by providing independent power supplies to each pump, or modelled, by adding an element for loss of power supply, in series with the system Reliability Block Diagram. In process plant, perhaps the systems which require the greatest care to be taken about common cause failure are emergency safeguard systems, failure of which on demand can have serious safety implications.

A number of ways of modelling unforeseen Common Cause Failure modes have been developed, two of which are presented briefly below. It must be noted that, since these failure modes are indeterminate, and little failure rate data is available because such events are comparatively rare, the use of these methods and their associated failure rates is somewhat subjective.

(a) *Additive cut-off*

Once the probability of failure of a system has been calculated assuming that failures are independent (that is, as described in previous chapters) an allowance can be added for Common Cause Failure. This, in effect, limits the reliability which a system can achieve. Thus, the true probability of system failure is given by

$$F_{\text{system}} = F_{\text{calculated}} + F_{\text{cut-off}}.$$

Values for cut-off probability are estimated according to the amount of diversity built into the system. For a simple redundancy safeguard/standby system, in which each item or subsystem in parallel is identical, a value of $F_{\text{cut-off}}$ of approximately 10^{-3} failures per demand has been proposed. It has been claimed that, by arranging that each subsystem comprises different manufacturers' equipment and/or different operating methods, the cut-off value could be progressively reduced to a minimum of approximately 10^{-6} failures per demand.

Example 15.4

A process requires a constant supply of cooling water. An auxiliary cooling water system is provided to supply water to the process if the main cooling water supply fails. The auxiliary cooling water system consists of four pumps in parallel, two driven by turbines and two by electric motors. Adequate cooling water is maintained if any one pump starts on demand. What is the probability that the auxiliary cooling water system will fail on demand?

Figure 15.3 shows a simplified schematic diagram of the auxiliary cooling water system. Previously calculated 'failure on demand' rates for each pump are shown (this calculation is a highly simplified extract from a much more detailed analysis of a real system).

The system is analysed in three stages, as follows.

(1) *Electrically driven pumps*
The two electrically-driven pumps may be considered as a sub-system of two pumps in parallel, that is, simple redundancy, but with some diversity of water supplies, so that a cut-off failure rate of 10^{-4} failures/demand may be assumed.

$$F_{\text{calculated}} = (4 \times 10^{-3}) \times (4 \times 10^{-3}) = 1.6 \times 10^{-5}$$

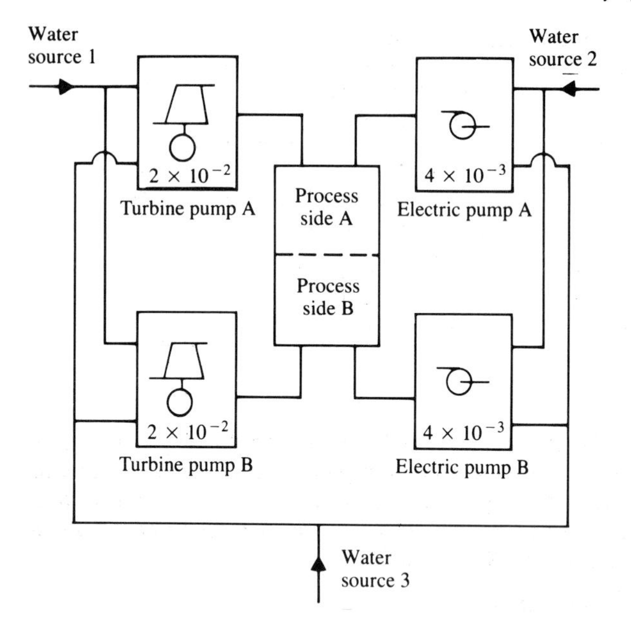

Fig. 15.3. Auxiliary cooling water system

so

$$F_{\text{sub-system}} = 1.6 \times 10^{-5} + 10^{-4}$$
$$= 1.16 \times 10^{-4} \text{ failures/demand}$$

(2) *Turbine driven pumps*
Similarly the two turbine driven pumps may be considered as a pair, with a cut-off failure rate of 10^{-4} failures/demand.

$$F_{\text{calculated}} = (2 \times 10^{-2}) \times (2 \times 10^{-2}) = 4 \times 10^{-4}$$

so

$$F_{\text{sub-system}} = 4 \times 10^{-4} + 10^{-4}$$
$$= 5 \times 10^{-4} \text{ failures/demand}$$

(3) *Auxiliary cooling water system*
The system fails only if both sub-systems fail, so

$$F_{\text{calculated}} = (5 \times 10^{-4}) \times (1.16 \times 10^{-4})$$
$$= 0.6 \times 10^{-7} \text{ failures/demand}$$

Assuming that the turbine driven subsystem and the electrically driven subsystem are so diverse that the minimum cut-off failure probability of 10^{-6} can be applied, the overall probability of failure on demand for the system is

$$F_{\text{system}} = 0.6 \times 10^{-7} + 10^{-6}$$
$$= 10^{-6} \text{ failures/demand}$$

Thus the overall system failure probability is dominated by the somewhat arbitrary figure for common cause failure. This is often the case for systems in which the lowest possible failure probabilities are demanded (that is, the highest levels of reliability are required).

(b) The beta factor method

An alternative method of allowing for common cause failures (CCF) is to treat the CCF not as an additional failure, as in the additive cut-off method, but as a percentage of the overall failure probability of each component within a redundant system. This is the beta factor method, so called because the CCF percentage of the overall failure probability is denoted by β.

Thus if a component has an overall failure probability, F

its common cause failure probability,
$$F_C = \beta F$$
its independent failure probability,
$$F_{IND} = (1 - \beta)F$$
overall failure probability,
$$F = F_{IND} + F_C$$
$$= (1 - \beta)F + \beta F$$

If a number of identical components are arranged in active-parallel, each has a common cause failure probability of βF, but a common cause failure can be expected to affect each of them virtually simultaneously. Consequently the system may be analysed as an active parallel group with a probability of failure, $F_{S.IND}$, due to independent failures in series with a probability of failure F_C due to common cause. That is

$$F_S = F_{S.IND} + F_C - F_{S.IND} \cdot F_C \quad \text{(from equation (9.3))}$$
$$= 1 - (1 - F_{S.IND})(1 - F_C) \quad (15.3)$$

For the simplest case of two identical components in parallel, system failure only if both fail, the system independent failure probability is

$$F_{S.IND} = (F_{IND})^2 \quad \text{(from equation (9.16))}$$
$$= (1 - \beta)^2 F^2$$

As explained

$$F_C = \beta F$$

Therefore

$$F_S = 1 - \{1 - (1 - \beta)^2 F^2\}(1 - \beta F)$$

Equation (15.3) is applicable generally, and may be used similarly to modify failure probability relationships, such as those in Table 10.2, to make an allowance for common cause failure.

Returning to the simplest case of two identical components in parallel, if the system is required to run for a given time, t, and its components both have constant mean failure rate, λ, and constant CCF percentage, β, the overall system failure probability is

$$F_S = 1 - \{1 - (1 - e^{-(1-\beta)\lambda t})^2\}e^{-\beta\lambda t}$$
$$= 1 - 2e^{-\lambda t} + e^{-(2-\beta)\lambda t} \quad (15.4)$$

The same approach may be used to obtain expressions of system failure probability for other parallel systems. A selection of the most common of these is given in Table 15.2, together with the equivalent expressions of failure probability for systems without Common Cause Failure (derived from equation (10.5)).

The main problem of using the beta factor method for new plant is the difficulty of estimating the beta factor before the plant is run. Assessments of running plant in a range of industries have derived beta factors mainly in

Table 15.2. Failure probabilities for systems with and without CCF

Redundancy level	*Probability of failure with CCF*	*Probability of failure without CCF*
1002*	$1 - 2e^{-\lambda t} + e^{-(2-\beta)\lambda t}$	$1 - 2e^{-\lambda t} + e^{-2\lambda t}$
1003	$1 - 3e^{-\lambda t} + 3e^{-(2-\beta)\lambda t} - e^{-(3-2\beta)\lambda t}$	$1 - 3e^{-\lambda t} + 3e^{-2\lambda t} - e^{-3\lambda t}$
2003	$1 - 3e^{-(2-\beta)\lambda t} + 2e^{-(3-2\beta)\lambda t}$	$1 - 3e^{-2\lambda t} + 2e^{-3\lambda t}$
2004	$1 - 6e^{-(2-\beta)\lambda t} + 8e^{-(3-2\beta)\lambda t} - 3e^{-(4-3\beta)\lambda t}$	$1 - 6e^{-2\lambda t} + 8e^{-3\lambda t} - 3e^{-4\lambda t}$
3004	$1 - 4e^{-(3-2\beta)\lambda t} + 3e^{-(4-3\beta)\lambda t}$	$1 - 4e^{-3\lambda t} + 3e^{-4\lambda t}$

* 1002 denotes 1-out-of-2, and so on.

the range 0.1–0.3. It is probably best, when assessing a new design, to carry out a series of calculations with different beta factors in order to check the sensitivity of the system reliability to Common Cause Failures.

Example 15.5

Two identical vacuum pumps are normally both working, but an adequate vacuum will be maintained if only one is working. The pumps have a constant mean failure rate of 10^{-5}/hr. Calculate the failure probability for a run of 1000 hours for a range of beta factors between 0 and 1.

$$\lambda t = 0.00001 \times 1000 = 0.01$$

$$e^{-\lambda t} = 0.9900$$

Table 15.3 displays a range of results obtained by using equation (15.4).

Table 15.3. Results for Example 15.5

β	F_s	*Comments*
0	9.90×10^{-5}	No common cause failures
0.05	5.89×10^{-4}	
0.1	1.08×10^{-3}	
0.2	2.06×10^{-3}	
0.3	3.04×10^{-3}	
1.0	9.95×10^{-3}	All common cause failures

15.5 Markov analysis of repairable systems

In the simple models of Chapter 9 it was assumed either that there was no repair of a failed unit, or that testing/repair time was negligible, or that an average unit availability could be assessed (which assumes implicitly that the repair process can be accounted for). We may, however, wish to assess the effect of a specified repair capability on system availability or reliability (survival probability) and, hence, wish to model a repair process explicitly.

Provided that average unit failure rates can be assumed constant (that is, negative exponential distribution of times-to-failure), and that the same applies to average unit repair rates (that is, negative exponential distribution of times-to-repair), then a technique known as *Markov Chain Theory* can be used to evaluate average availabilities, or reliabilities, of at least the more straightforward models of repairable systems. The analysis involves matrix algebra solution of simultaneous differential equations describing the evolution of system state probabilities, and is described in some detail in references (**19**) and (**20**).

Example 15.6

Consider the simple active parallel system of Fig. 15.4(a). Assuming that a pump is either fully operational

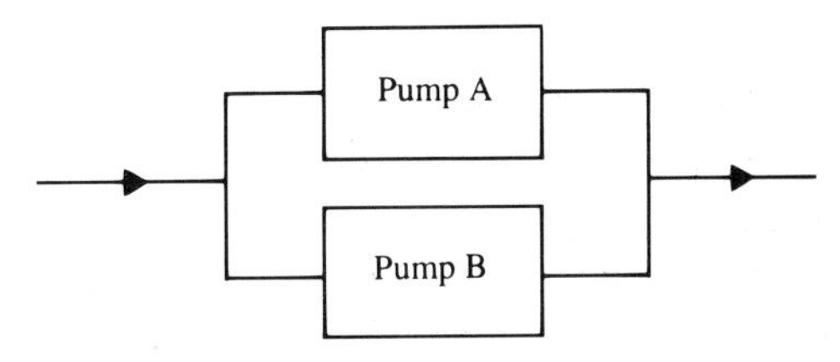

(a) Reliability block diagram

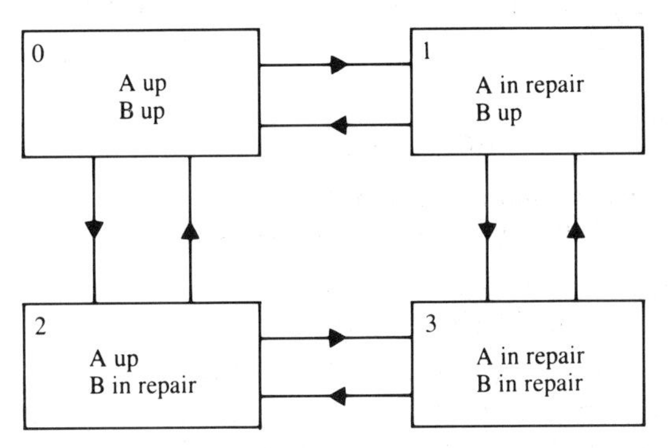

(b) State-space diagram

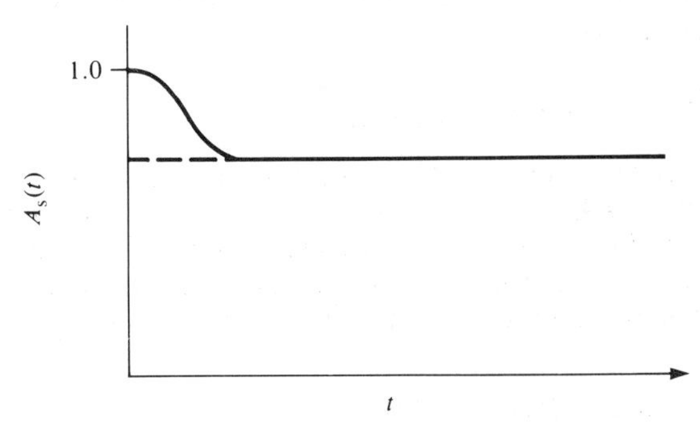

(c) Plot of expected availability with time

Fig. 15.4. Simple system with repair

or under repair (that is, in this case, there is no queuing for repair), then the system can be in one of four states as indicated in the *state-space diagram*, Fig. 15.4(b), which also shows the possible transitions between the states. Markov analysis evaluates:

(1) the probability that at any time, t, the system will be in a given state, given that it was in state 0 at $t = 0$; hence,

(2) the average, or expected, system availability at any time, t.

The result is as in Fig. 15.4(c), which illustrates that, not unexpectedly, the system availability is high early on (there being a very low probability at that time that the system could have evolved to state 3) and that eventually it settles down to a constant average, steady-state. For this example, if:

(1) the pumps are identical and of mean failure rate, λ (that is, MTTF = $1/\lambda$);

(2) the mean repair rate per unit is μ (that is, MTTR = $1/\mu$);

(3) there is no waiting for repair (that is, there is always a repair team available);

then the steady-state system availability is

$$A_s = \frac{\mu^2 + 2\mu\lambda}{(\mu + \lambda)^2} \qquad (15.5)$$

Thus if

$$\lambda = 0.03/\text{day} \quad (\text{MTTF} = 33.3 \text{ days})$$

$$\mu = 0.4/\text{day} \quad (\text{MTTR} = 2.5 \text{ days})$$

$$A_s = \frac{0.4^2 + 2 \times 0.4 \times 0.03}{0.43^2}$$

$$= 0.995 \text{ (or 99.5 per cent)}$$

Solutions of some of the simpler cases are given in Table 15.4; note that some of the cases have a limiting repair capacity; that is, waiting for repair is a possibility. Computer programs are available for solving more complicated cases. For these, it is important that the state-space diagram models the possible states, transitions, and transition rates as closely as possible.

15.6 Monte Carlo simulation

When we wish to investigate more complicated operation and failure patterns than have been described above, or detailed aspects of equipment repair, such as spares holding, delays before repair can start, or priorities where there are repair resource limitations, the mathematical analysis can be extremely difficult or impossible to solve.

Monte Carlo simulation is a method which can be used to bypass the complex mathematics of an analytical solution. It can only be used effectively with a computer and was, at one time, considered too expensive to use routinely. However, with the advent of the microcomputer and the appearance of Monte Carlo simulation programs on the market, the technique is becoming more accessible.

The technique is to generate a computer model of the system to be investigated, and then to simulate the operation of the system for a predetermined period, during which random failures and repairs can occur to the components of the system. The operational states which the system takes up as a result of each failure or repair (or other event) are logged, and from the percentages of time spent in each state, the overall system availability can be calculated and other useful information may be inferred.

(a) The computer model

The computer model is usually based on a reliability block diagram of the system, and controlled by a set of rules which specify exactly the model's response to each type of event which can occur. Each block in the reliabil-

Table 15.4. Markov solutions for some simple systems

System	*Steady-state availability*	*Approximate reliability, $R(t)$, if $\lambda \leqslant 10\mu$*
2 units in active parallel, no waiting for repair	$\dfrac{\mu^2 + 2\mu\lambda}{(\mu + \lambda)^2}$	
2 units in active parallel, only one repairable at a time	$\dfrac{\mu^2 + 2\mu\lambda}{\mu^2 + 2\mu\lambda + 2\lambda^2}$	$e^{[-1/2\{(3\lambda+\mu)-(\mu^3+6\mu\lambda+\lambda^3)^{1/2}\}t]}$
2 units in standby parallel, no waiting for repair, no standby failures	$\dfrac{2\mu^2 + 2\mu\lambda}{2\mu^2 + 2\mu\lambda + \lambda^2}$	
2 units in standby parallel, only one repairable at a time, no standby failures	$\dfrac{\mu^2 + \mu\lambda}{\mu^2 + \mu\lambda + \lambda^2}$	$e^{[-1/2\{(2\lambda+\mu)-(\mu^2+4\mu\lambda)^{1/2}\}t]}$

Assume identical units, unit failure rate = λ, unit repair rate = μ, and no failures on standby.

ity diagram (which can represent a component or group of components, or even one aspect of a single component), can be assigned an individual failure rate, repair rate, and number of spares available. The set of rules specifies such details as which blocks (components) have to be 'taken out of service' if a block fails, which other blocks have to be 'put into service', which repair strategy is put into practice, for example, whether the 'component' is to be repaired *in situ* or changed for a spare, and whether an exchanged component should be scrapped or refurbished, and the effect of the block failure on the percentage throughput of the whole system.

For example, if a system model includes two 100 per cent throughput feed pumps in parallel, and each pump set is treated as four components in series (that is, high pressure pump, booster pump, electric motor, and balance of plant) the rules will probably require that if the electric motor of the running pump fails, the remaining components of the running pump are shut down and the standby pump is started. At the same time the rules will initiate the repair of the failed motor, perhaps replacing it with a spare if one is available and to do so would be quicker than repairing it *in situ*. In this case, if the standby pump is permitted to start without failure, the availability of the whole system model is unaffected. Further rules are required to determine whether the failed pump is returned to service as soon as it has been repaired or remains as standby until the other pump fails.

(b) *Simulation of events*

When the model has been set up in the computer, it is run by simulating sequences of events (failures, repairs, consumption of spares, and so on) which occur independently to each component of the model. For the model to mimic the real system properly, the events must occur at intervals related to those which could occur in the real system. If, for instance, a real component has a mean time to failure of 3000 hours and the distribution of the times to failure is known (say, exponential distribution), then the set of times to failure which occur to that component in the model must be drawn from that distribution and must represent a mean time to failure which approximates to 3000 hours. The computer must, therefore, generate an independent series of random times for each component parameter, to suit the specified mean times and distributions. Unlike Markov analysis, the Monte Carlo simulation method is not restricted to use of the exponential distribution, but can simulate times drawn from any distribution which seems appropriate (for example, Weibull, log-normal, rectangular, and so on).

When the simulation is set in motion, failures, repairs, and such like occur to the components of the model at the times specified by the set of random time series and controlled by the model's set of rules. The state of each component is logged after either each unit time interval or each change of state of any component.

The length of the simulation may be expressed in terms of the time for a specified number of failures to occur in the model (perhaps several thousand) or the time for a number of cycles of specified length, for example, the time between major overhauls or the complete life cycle. Because the model can run so much faster in the computer than a system in real time, very long runs are possible. In general, the longer the run, the closer the sets of random time series will approximate to the desired distributions, and the closer the overall system availability will approach a steady-state value.

(c) *Results*

At the end of the simulation, the program totals the time spent by each component in its running, standby, and failed states, and the time spent by the overall system in all its possible states, from which the overall system availability can be calculated. Other subsidiary information can also be obtained, such as the number of times each component failed, or standby plant was called to start, the number of times components were exchanged for spares, and whether there would have been advantage in having more spares available.

(d) *Comments*

The above description is not of a specific computer program but shows, rather, how a Monte Carlo simulation can be carried out and some of the features which may be found in programs. Not all commercially-available programs will have all the features in this description, but may have some additional features.

Monte Carlo simulation is a powerful technique, which is capable of producing an answer to any problem posed by the reliability engineer, subject only to the ingenuity of the computer programmer (and the cost of his time). However, it is not an analytical technique, and the more complex the problem, the more difficult it is to check if the program has been written correctly and, therefore, if the result can be relied upon. Also, the more components and rules in the model, the longer it will have to run in order to achieve a steady-state result. While a substantial simulation may be carried out in seconds on a large computer, it can take several hours on a microcomputer, so that there will usually have to be a compromise between computing time and costs and the desire to ensure that the steady-state solution has been reached. This is particularly so when several runs of the same model are required, with different sets of random time series to check consistency, or when the model is run many times with different component parameter values or configurations when searching for an optimum design for a real system.

PART FIVE

Collection and processing of reliability data

CHAPTER 16. INTRODUCTION TO DATA COLLECTION

16.1 Background and need

Parts Three and Four have shown that, for the reliability assessment of new process plant designs, information on past operational experience is required. Chapter 9 has shown that simplified reliability models of the new plant system, in the form of reliability block diagrams and logic block diagrams, are used to identify sub-systems and components for which adequate reliability data are required. Clearly, high quality failure and repair data are required for these models if realistic predictions of system reliability characteristics are to be achieved.

Reliability data can be provided from two basic sources.

- Published data available internationally from a number of sources.
- In-house records of plant operating experience collected by companies for their own internal use.

Whatever the source, data for mechanical equipment and components is a particularly difficult problem area and may be subject to considerable uncertainty. Part Five reviews the main problem areas and outlines the processes involved in selecting and modifying the basic failure information to obtain realistic data for reliability predictions.

16.2 Quality aspects of mechanical reliability data

Because of the uncertainties mentioned above, when using reliability data for mechanical equipment and components for the prediction of reliability for new process plant systems, it is always better to use in-house data, as the source of the data and the environment in which it was collected are fully known, and it is more likely to be relevant to the needs of the new plant.

To obtain high quality reliability data, the collection and analysis of failure statistics for each component and equipment type from a representative population operating under identical conditions is needed. This is obviously an ideal situation, and never possible in practice for mechanical equipment.

In the electronics industry, where component parts, such as resistors and capacitors, are relatively cheap, comprehensive programmes of reliability testing have been carried out by a number of organizations. The objective has been to identify failure characteristics of different standard component types over a wide range of operational and environmental conditions. The results of the tests have been analysed, compiled into sets of tables and graphs for different stress conditions, and published periodically as generic data handbooks. These handbooks provide recommended component failure rates from which the reliability of electronic equipment can be assessed.

For mechanical systems the utilization of a wide range of components, many non-standard, their size, and the costs involved, are the major factors which inhibit the scope for such a reliability testing programme. Thus, the failure data published in the open literature and available from various reliability data banks are generally assembled at equipment level from the analysis of failures in the field.

When reliability assessments of mechanical systems are carried out, these component and equipment reliability data need to be adjusted to reflect the expected operational and environmental conditions, usually without any of the stress and load information that is available for electronic components. In consequence, there is a much higher degree of uncertainty associated with the reliability data available for mechanical system assessments, and more time and experience are required to derive meaningful figures for each study from the published data available.

To limit uncertainties, in-house field data collection should be carried out under controlled conditions where each failure is recorded as it occurs. For these failure events, critical information such as failure modes, failure causes, and operational and environmental conditions, can generally be correctly identified and recorded for known equipment populations. Examples of such reliability data collection schemes under controlled conditions are, however, quite rare.

The studies published by Électricité de France on pump and valve data collection in nuclear power stations provide some indication of the large investment of resources needed to establish and support such a scheme. Consequently, only in very exceptional circumstances are reliability data collected on operating process plant for mechanical equipment at this detailed level, despite the growth of computerized systems. In the majority of cases, historical data, obtained from recent maintenance records, test reports, and operational logbooks provide the basis for mechanical reliability data analysis.

The uncertainties arising from the analysis of historical maintenance records can be considerable. Clearly, the information will have been collected to support maintenance planning. The emphasis will thus be on the repair actions rather than the cause and affect of each failure. Translating maintenance records into the population and failure statistics needed for reliability data analysis always takes time and considerable engineering judgement. Additional information from test reports, operational logbooks, and manufacturers catalogues may also be required to complete the picture and reduce the uncertainties.

Other causes of uncertainty arise from the variations in operational, environmental, and repair conditions associated with individual items of equipment. The plant will seldom be new, and equipment age and previous repair history will frequently be unclear. Sample populations may also be small (that is, the number of similar items of equipment under study may be small).

Table 16.1. Advantages and disadvantages of using generic published data or collecting in-house data

Data source	*Advantages*	*Disadvantages*
Generic – published	Easy to obtain	Non-specific boundaries and assumptions unknown
Generic – applied	Can be produced relatively quickly	Requires detailed knowledge of proposed application and factors influencing reliability
Collected – maintenance data	Uses existing records	May not have data in the correct form or detail required
Collected – formal scheme	Very detailed reliability parameters can be obtained on specific items and in correct form	Long term project in order to collect sufficient experience

16.3 Measuring in-service reliability

The manner and detail in which reliability data is collected depends on a number of factors. These will normally be organizational constraints; for example costs, manpower available, availability of computerised systems, the uses to which the data will be put, and so on, are of prime importance. The uses of the collected data will include one or more of the following:

(1) Detection of in-service problems and identification of causes in a quantitative form.

(2) Monitoring maintenance and impact of operational changes.

(3) Evaluation of plant reliability and availability.

(4) Creating and/or updating a reliability data bank, including system, sub-system, and component failure rates and modes. Also, details on maintenance hours for each category of equipment repaired or maintained.

(5) Feeding the operational reliability data into a central library for future reliability predictions, and comparing achieved plant reliability with design predictions or to assess performance against specified reliability.

(6) Establishing reliability performance criteria for future plant and equipment.

16.4 Failure data collection and processing

Failure data collection is essential to both operations, engineering, and specialist reliability staff. The maintenance engineer requires systematic feedback of failures occurring on plant to identify those items causing excessive down time, maintenance effort, and material costs. Also, quantified operational failure data provides a fast identification of in-service problems and assists operations staff with the interface with suppliers to respond to plant operational problems. The reliability engineer needs a source of accurate, rated data to enable designs to be assessed and optimized for future plant with a high degree of confidence in the assessment.

However, the raw data needs to be subjected to a process of modification and analysed before it can be of direct use for reliability prediction.

The main advantages and disadvantages of using generic published data or collecting in-house data are summarized in Table 16.1.

CHAPTER 17. DATA REQUIREMENTS AND RELIABILITY PARAMETERS

17.1 General

Reliability prediction for the design of new process systems is carried out in three main areas.

(1) Safety studies:
- to evaluate reliability performance against risk targets;
- to identify weak areas in the design;
- to provide a sound basis for safety testing programmes.

(2) Plant availability:
- to evaluate outage times and production losses against economic targets;
- to identify critical sub-systems and components;
- to determine the need for redundant or standby equipment.

(3) Maintainability studies:
- to determine the most effective maintenance strategy;
- to optimise maintenance routines, spares holdings, manning levels, and so on;
- to assess the need for condition monitoring.

The reliability parameters which are fed into mechanical system models are identical in most respects to those employed for electronic systems, however, the emphasis tends to be different – for example the repair process may be more significant in mechanical system studies. Maintainability aspects will generally be more complex and more time will need to be devoted to evaluating the data to ensure that failure modes and restoration actions which affect operational reliability are properly identified and quantified.

The main reliability data required in predictive safety, availability, and maintenance studies are shown in Fig. 17.1. These parameters are defined in BS 4778:1979 (**21**), but some comments are worthwhile here on the parameters featured in published lists of mechanical equipment failure rates and reliability data banks.

These data are predominantly concerned with overall failure rate statistics but, although it is less frequently available, it is also desirable to have information on:

failure mode proportions;

spurious trip rate;

failure on demand rate;

mean time to restore.

17.2 Reliability parameters

The main parameters required for system reliability assessment are discussed below.

(a) Overall failure rate

Failure rate is the most frequently quoted reliability parameter for mechanical components. The overall failure rate is assumed to relate to the useful-life (constant failure rate) phase and to include all failures – those which cause changes in the operating state and those which do not. The latter category covers failures which result in some degradation in performance and equipment deterioration rectified during routine maintenance.

For random failures during normal operation the overall failure rate stated frequently assumes that outages due to repair are negligible. The failure rate is then

$$\lambda = \frac{\text{No. of failures in interval}}{\text{Component population} \times \text{calendar time interval}}$$

$$= \frac{n}{N \times T}$$

The units quoted are normally failures/10^6 hours or failures/year.

In some reliability data banks an *operating-time* failure rate may also be given.

The probability of no failures (reliability) for a time period, t, can be calculated from the expression

$$R(t) = e^{-\lambda t} \qquad (17.1)$$

where

$R(t)$ = probability of successful operation for a period of time, t

λ = failure rate

e = base of natural logarithms

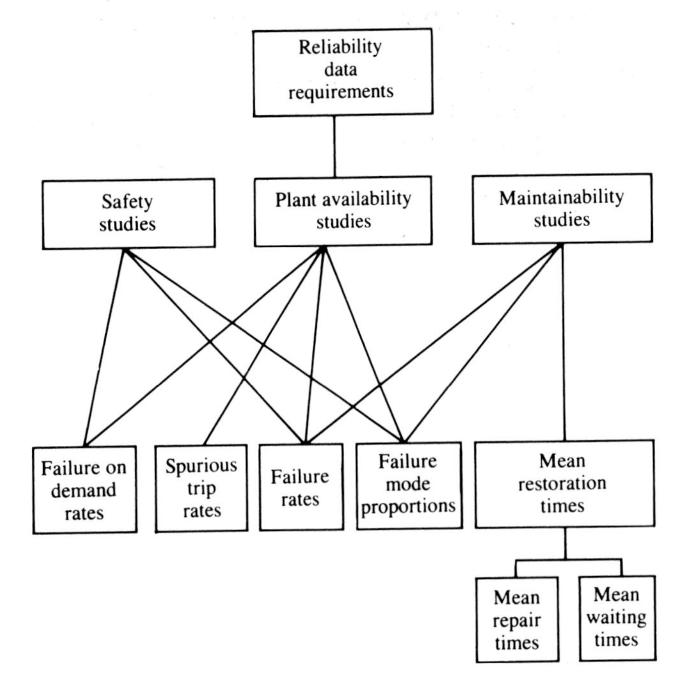

Fig. 17.1. Reliability data requirements

(b) Failure rate proportions

The overall failure rate of an equipment or component class is estimated by aggregating failures in a number of different failure modes. Each of these component failure modes must be considered to determine its effect on the system being assessed. Only those failure modes which move the system into an undesired operating state (for example, 'dangerous') are employed when calculating the reliability of the system. Hence, the overall failure rate must be adjusted to weed out the non-significant failure modes before carrying out the system reliability calculation.

If only modes A and B of component X result in 'failure' of the system then

$$\lambda_x = \{p(A) + p(B)\} \cdot \lambda_0 \tag{17.2}$$

where

λ_x = failure rate of component X which results in system failure
$p(A)$ = proportion of failures in mode A
$p(B)$ = proportion of failures in mode B
λ_0 = overall failure rate of component X

The reliability of a series system comprising components X, Y, and Z can then be calculated from the expression

$$R(t) = e^{-\lambda_x t} \cdot e^{-\lambda_y t} \cdot e^{-\lambda_z t} = e^{-(\lambda_x + \lambda_y + \lambda_z)t} \tag{17.3}$$

where the λs have been adjusted to reflect those failure modes which result in system failure.

(c) Spurious Trip Rate

The Spurious Trip Rate is the frequency of false alarms which result in system shutdown. Generally, the false alarm is caused by abnormal stimulation of protection or control instrumentation and not by component failure. An example is sunlight reflecting from the sea on to one or more ultra-violet flame detectors in the fire-detection system of an offshore platform.

Spurious Trip Rates at component level are difficult to derive and, thus, are seldom featured in reliability data banks. If it is necessary to estimate the number of shutdowns or the outage time due to false-alarms during a reliability study, data will generally be collected at system or major sub-system level, where logbooks are available recording every change in system operating state. At this level the Spurious Trip Rate can be defined as

$$\lambda_s = \frac{\text{No. of spurious trips}}{\text{Total operating time}}$$

The rate will generally be quoted in spurious trips/year.

The probability of trip-free operation for a period of time can be calculated from the expression

$$R(t) = e^{-\lambda_s t} \tag{17.4}$$

The probability of one, or more trips in the interval, t, is the complement of $R(t)$, that is

$$F(t) = 1 - e^{-\lambda_s t} \tag{17.5}$$

(d) Failure-on-demand rate

Certain items of equipment on process plants are only called on to operate when required. Examples are pumps on standby duty, emergency power generation, and protective instrumentation. Failures which occur when the equipment is in the dormant or standby state will only become apparent during a test or when a demand is made on the system.

Faults of this type are known as 'unrevealed' failures. Since test intervals will obviously be set to ensure that the probability of failure-on-demand is small, a reasonable approximation of the failure-on-demand rate can be obtained from test records. The failure-on-demand rate (or unrevealed failure rate) is then

$$\lambda_d = \frac{\text{No. of failed tests}}{\text{No. of tests} \times \text{test interval}} = \frac{n_f}{N_\tau \times T}$$

The units normally quoted are failures/year or failures/10^6 hours. In certain cases failures/demand may be given.

Assuming a random arrival of demands on the system the mean of a large number of demands will be at $T/2$, that is, half way through the test interval. The mean probability of failure on demand or Fractional Dead Time of the system is thus

$$\text{FDT} = \frac{\lambda_d T}{2} \tag{17.6}$$

where

λ_d = failure-on-demand rate
T = test interval

If two parallel systems are fitted the probability of both systems being in the failed state on demand:

$$\text{FDT} = \frac{\lambda_d^2 T^2}{3} \tag{17.7}$$

Formulae for calculating the fractional dead times for higher order parallel systems and majority-voting systems (for example, 2 out of 3 dormant systems being in the failed state on demand) are derived in Part Three, Chapter 10.

(e) Mean time to restore

With certain equipment the time to restore operation after failure can be significant. In these situations the Availability function gives a better representation of system reliability. For steady-state conditions, Availability is the ratio of operating time to total time; hence

$$A = \frac{\text{MTBF}}{\text{MTBF} + \text{MTTR}} \tag{17.8}$$

where

MTBF = mean time between failures = $1/\lambda$
MTTR = mean time to restore operation

Its complement, the Unavailability of the system

$$U = \frac{\text{MTTR}}{\text{MTBF} + \text{MTTR}} \qquad (17.9)$$

Since $A + U = 1$ it is clear that Availability, like Reliability is a probability; effectively the probability that the system is operating at some arbitrary time, t, in the interval between major overhauls.

The mean time to restore is made up of two components; the Mean Repair Time (MRT) and the Mean Waiting Time (MWT). The units normally quoted are hours. In most cases only repair times will be recorded in maintenance records and an estimate of the Mean Waiting Time will need to be made to add to the Mean Repair Time to obtain the mean time to restore operation. Times which need to be considered when estimating mean waiting time include time from failure to the start of repair and time from completion of repair to return to service of the repaired item.

In certain circumstances a detailed Maintainability Analysis may be performed to estimate the mean time to restore. The method, described in MilHdbk 472 – *Maintainability analysis* (**22**) is quite involved and outside the scope of this book.

17.3 Class characteristics

A system reliability assessment involves the decomposition of the system to a unit or component level at which suitable failure data exist. Failure modes which cause a change in the system operating state are then identified and failure probabilities calculated from generic reliability data for application in the system model. It is essential that these generic data relate to unit or component classes which reflect similar design and functional characteristics.

Quite detailed classifications (taxonomies) have therefore been developed, notably by the nuclear industry, to describe component classes. One such classification employs a 14 digit code spread over 7 fields. For mechanical valves, inventory information is coded into the 7 fields as follows.

Field	Description	Digits	Code	Meaning
Field 1	Main component class	5 digits	92300	Mechanical valve (including actuator)
Field 2	Component type	2 digits	36	Globe
Field 3	Method of actuation	2 digits	66	Pneumatic diaphragm/cylinder
Field 4	Trim materials	1 digit	4	Nickel based alloys
Field 5	Function	2 digits	15	Flow control
Field 6	Pressure range	1 digit	7	1000–3000 lb/in^2
Field 7	Temperature	1 digit	5	100–250°C

The code 92300.36.66.4.15.7.6. thus identifies the component as a Mechanical valve (including actuator), globe type, pneumatic diaphragm/cylinder actuated, with nickel based alloy trim, used for flow control in the pressure range 1000–3000 lb/in^2, and temperature range 100–250°C. This coded description identifies a very specific class of mechanical valve, and hence, sample populations are likely to be very small if, in fact, any exist at all within the data collected. Inevitably a restricted definition (for example, 92300. . . . .15 – Mechanical valve (including actuator) used as a flow control valve) will need to be selected to give a sufficiently large sample population to warrant statistical analysis.

These restricted component class descriptions are generally sufficient to derive the global failure rates employed in initial reliability studies. In consequence most published lists of generic reliability data are com-

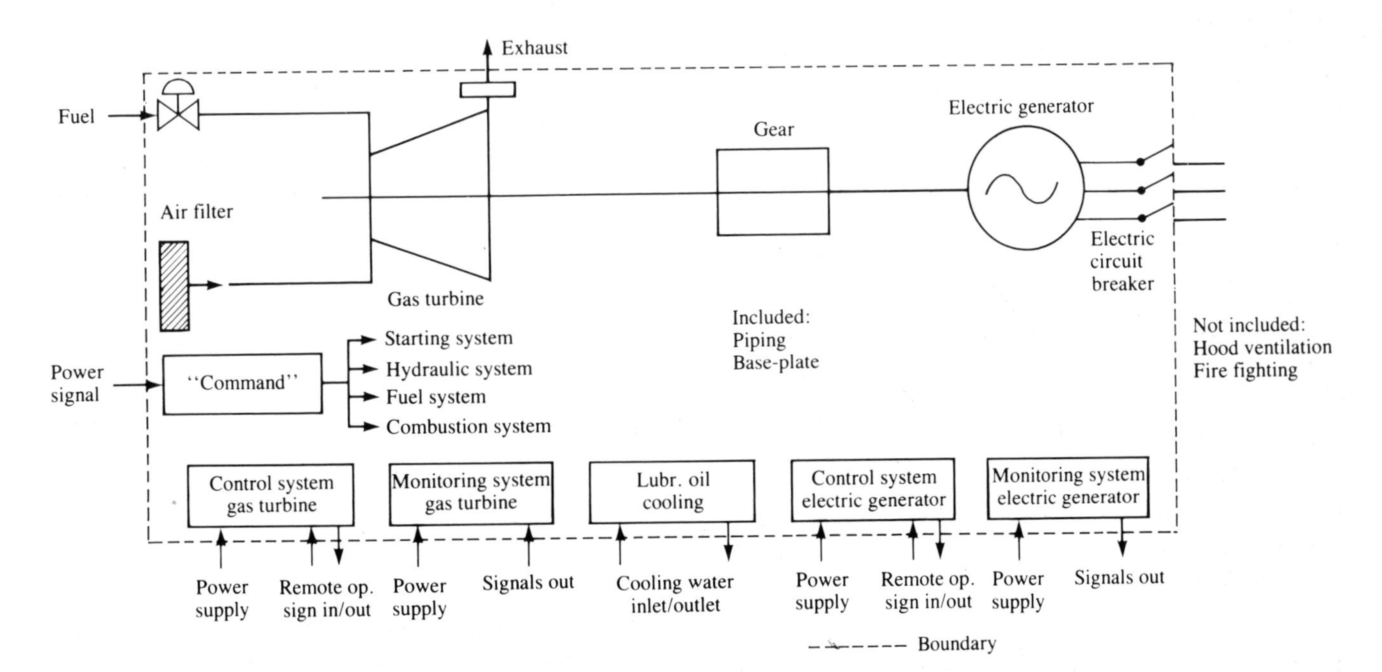

Fig. 17.2. Boundary specification – electric power generating sets

Table 17.1. Reliability data – electric power generation sets

Taxonomy No.	*Item*
3.1.1.1.1	*Electrical system* *Power generation* *Main power generators (incl. drive), turbine driven, industrial* *Less than 7 MW* *Gas fuel*

Population	*Samples*	*Aggregated time in service (10^6 hours)*		*No. of demands*
		Calendar time	*Operational time*	
4	2	1840	0624	

Failure mode	*No. of failures*	*Failure rate (per 10^6 hrs)* Lower	Mean	Upper	*Active repair (hours)*	*Repair (manhours)* Min.	Mean	Max.
Critical	6 6	0 0	43 110	130 300		—	8	—
Failed to start	6 6	0 0	43 110	130 300		1	8	26
Degraded	15 15	40 140	71 230	120 360		—	14	—
Failed to start and supply load within specified time	9 9	19 67	43 140	79 240		5	17	46
Improper operation	6 6	10 36	28 91	60 180		3	9	20
Incipient	46 46	170 530	230 700	300 900		—	13	—
Overheated	16 16	44 150	76 240	120 370		1	8.5	20
Faulty indication	4 4	0 18	26 60	78 140		2	12	24
r/min hunting	3 3	2.4 10	14 45	40 120		2	17	41
Vibration	2 2	0.55 3.8	9.5 30	33 99		36	42	48
Unknown	21 21	62 210	100 320	150 460		1	14	64
Unknown	4 4	4.7 18	19 60	47 140		—	16	—
Failed	4 4	4.7 18	19 60	47 140		4	16	36
All modes	71 71	290 880	360 1100	440 1300	—	—	13	—

Comments

piled at a two-descriptor level (for example, pump, centrifugal). Where sensitivity analysis shows that the failure rate of a particular class of component is critical to system reliability, then more detailed study may be warranted. This is likely to include access to a reliability data bank where a more sophisticated classification such as the one described previously may be employed to obtain more component-specific data.

For critical components it is always necessary to identify the boundaries and conditions under which the failure data has been collected. In the mechanical valve example given previously the boundaries are well defined by the 14 digit code and additional information on the operating environment will also be given. However, when the data are collected on units rather than components, it is necessary to be quite specific about the components which are included within the unit boundary, the application, operating mode, environment, and so on, to ensure that uncertainties associated with the data are minimized. The application, operating mode, and environment may also be important.

An example of a unit-level boundary definition is shown in Fig. 17.2 for electric power generator sets. The source is the *OREDA–84 Handbook* (**14**). The reliability data associated with these units are shown in Table 17.1.

CHAPTER 18. GENERIC FAILURE RATE APPLICATION

As noted previously it is essential that representative data are input to mechanical reliability models. For initial studies carried out during the conceptual design stage of a project, global data, suitably adjusted to reflect the critical failure modes, operating duty, and environmental stress levels, will frequently be adequate. However, during detail design more information will become available as to the make, model, operating conditions, environment, and so on, of each item of equipment. Particularly for critical items in the system, it will then be worthwhile identifying and analysing the most relevant data available in the light of the principal stresses influencing equipment failure characteristics.

Failure rate prediction models are generally of the form

$$\lambda_{XA} = \lambda_0(K_1 \cdot K_2 \cdot \cdots \cdot K_n) \cdot p(A) \quad (18.1)$$

where

λ_{XA} = predicted failure rate for equipment X in failure mode A

λ_0 = overall failure rate for equipment types similar to X

K_i = stress factor for stress i

$p(A)$ = proportion of failures in mode A

For global data in conceptual design reliability studies, a simplified model is employed

$$\lambda_{XA} = \lambda_0 \cdot K_1 \cdot K_2 \cdot p(A) \quad (18.2)$$

where

K_1 = stress factor for the overall environment

K_2 = stress factor based on the component nominal rating

Tables 18.1, 18.2, and 18.3 show recommended failure rates for typical process equipment, and stress factors for environment and rating, respectively. The proportion of failures in the relevant failure mode $p(A)$ may be available from in-house or published information. Alternatively it may be estimated from engineering experience.

An example of failure rate prediction for a control valve and centrifugal pump operating in an offshore environment is shown in Table 18.4.

Confidence limits for the centrifugal pump are shown in the data source reference – adjusted for the increase in duty arising from operation in crude oil, an upper 95 per cent confidence limit of 360f/10^6h can be assumed. For the control valve, however, only a mean failure rate is given. Studies based on mean failure rates have been the subject of a number of investigations. The most recent by Snaith (**23**) confirmed previous results which indicate that approximately 60 per cent of all predictions studied lie within a factor of 2 of the mean and 90 per cent within a factor of 4. As a first approximation, therefore, the upper 95 per cent confidence limit for the predicted control valve failure derived in Table 18.4 is 45f/10^6h.

For deriving confidence limits where sample data are available, see section 7.3.

Table 18.1. Generic failure rates – process equipment

Equipment	*Failure rate (f/10^6 hours)*
Accelerometer	5
Actuator	36
Alarm bell	5
Air supply	8
Alarm system	24
Amplifier	18
Analyser – chemical	530
Battery	18
Battery charger	26
Blower	185
Boiler, steam	220
Busbar	17
Cable, power	3
Circuit breaker	0.3
Clutch	5
Compressor – air	51
Compressor – centrifugal – large	4000
Compressor – reciprocating – large	8000
Condenser, vapour	26
Connector, electric	0.4
Computer – Mainframe	6000
– Mini	400
– Micro	60
Contractor – Glycol	90
Controller temperature	30
pressure	12
flow	30
level	10
Converter, electronic	10
Conveyor	670
Coupling	17
Crane	2000
Detector – Gas	47
– H_2S	18
– Smoke	7
– UV	7
– Infra-red	12
Drier, small	45
Drier, large	3000
Fan	50
Filter, general	38
Filter, mechanical	130
Generator – industrial diesel	900
Generator – aero-derivative	1800
Heat exchanger – shell/tube (large)	60
– shell/tube (small)	20
Heat exchanger – Plate	30
Heater – electric	90
Indicator temperature	100
pressure	12
flow	30
level	20
Inverter	40

Continued

Table 18.1. *Continued*

Equipment	*Failure rate ($f/10^6$ hours)*
Joint, pipe	2
Motor – electric	8
Oscillator, electronic	26
Power supply	32
Pump – centrifugal – large	300
Pump – centrifugal – small	100
Pump – reciprocating	350
Pump – fire – engine driven	500
Recorder, temperature	110
Relay	1
Seal	4
Switch – electric	2
Switch, pressure	5
level	30
Separator	170
Scrubber	14
Tank	4
Timer	42
Transformer	5
Transmitter, temperature	10
pressure	14
flow	38
level	20
Turbine, steam	30
Valve – motorised	35
– pneumatic	60
– hydraulic	50
– deluge	20
– manual	10
– blowdown	30
– safety – relief	30
– master	50
– wing	25
– DHSV	40
Vessel – pressure	1

Table 18.2. Environmental stress factors (K_1)

General environmental conditions	K_1
Ideal, static conditions	0.1
Vibration free, controlled environment	0.5
General-purpose, ground-based	1.0
Ship, sheltered	1.5
Ship, exposed	2.0
Road	3.0
Rail	4.0
Air	10.0
Missile	100.0

Table 18.3. Rating stress factors (K_2)

Percentage of component nominal rating	K_2
140	4.0
120	2.0
100	1.0
80	0.6
60	0.3
40	0.2
20	0.1

Table 18.4. Failure rate prediction for control valve and centrifugal pump (offshore application)

Part No.	*Description*	*Overall failure rate*	*Reported failure modes*		*Data source**	*Relevant failure mode*	*Contribution to RFM $(PX) \times 100$*	*Environment (K_1)*	*Duty (K_2)*		*Predicted failure rate for RFM*
1	Control valve	$300f/10^6h$			(1)	Fail-open					
			Leaks	50%	(2)		10%	Ship-sheltered	Media – crude oil	1.125	$= 11.25f/10^6h$
			Actuator	30%			100%		Pressure – 500 lb/in²	6.75	
			Seize	10%			100%		Temperature – 40°C	2.25	
			Others	10%			50%			1.125	
								$K_1 = 1.5$	$K_2 = 0.5$		
2	Centrifugal pump (Multi-stage) 100HP	$370f/10^6h$ (Water)			(3)	No-flow					
			Critical	40%	(3)		100%	Offshore	Media – crude oil	225	$225f/10^6h$
			Degraded	13%			0%			0	
			Incipient	45%			0%			0	
			Unknown	2%			0%			0	
								$K_1 = 1$	$K_2 = 1.5$		

*(1) Table 8.1
(2) *The reliability of process control equipment* – T. R. Moss.
(3) *OREDA – Offshore Reliability Data*.

CHAPTER 19. SOURCES OF MECHANICAL RELIABILITY DATA

Table 18.1 provides a useful list of typical failure rates for process equipment – it has been developed over a number of years and is generally based on relatively large populations of equipment items. Similar lists have been published from time to time, for example by Rothbart (**24**), Green and Bourne (**25**), and Smith (**26**).

A number of reliability data handbooks have also been published which feature mechanical items. The most notable items are the *Non-Electronic Reliability Notebook* (**27**), covering mechanical and electrical components in military and commercial applications; the *IEEE Standard 500* (**28**), featuring mechanical, electrical and instrument equipment in nuclear power stations and the *OREDA Handbook* (**14**) which provides a range of detailed reliability and maintainability information for offshore mechanical, electrical, and instrument systems and equipment. Failure mode information, failure rates, and confidence limits are given in many cases or can be derived from the failure statistics provided in the handbooks. Typical ranges of failure rates for components, equipment, and systems are shown in Fig. 19.1.

Laboratory testing to obtain reliability data for mechanical equipment is seldom economic, so that most in-house data will be generated from field experience. Many organizations collect data that can be analysed to provide reliability information. These data exist in maintenance and test records, operational logbooks, and other technical information systems. Particularly in the present climate, where a wide range of computer systems are available for maintenance planning, there exists the basis for comprehensive in-house reliability data bases in most companies.

The problem with in-house data is that population sizes for important items are frequently quite small so that reasonable confidence in the data can only be obtained after relatively long periods of operation. Generic reliability data from one or more of the commercially available databanks can be useful to supplement or reinforce the failure rate estimates obtained from in-house sources.

For industrial equipment the Systems Reliability Service Data Bank (SYREL) operated by the National Centre of Systems Reliability (NCSR) provides the most comprehensive source of reliability data in the UK.

Maintainability data is generally considered to come within the orbit of reliability data banks, but until recently has received little attention. With the increased interest in availability of process plant it is clearly an area needing attention. In a study of Brent Gas Disposal System performance (**29**) for example, it was found

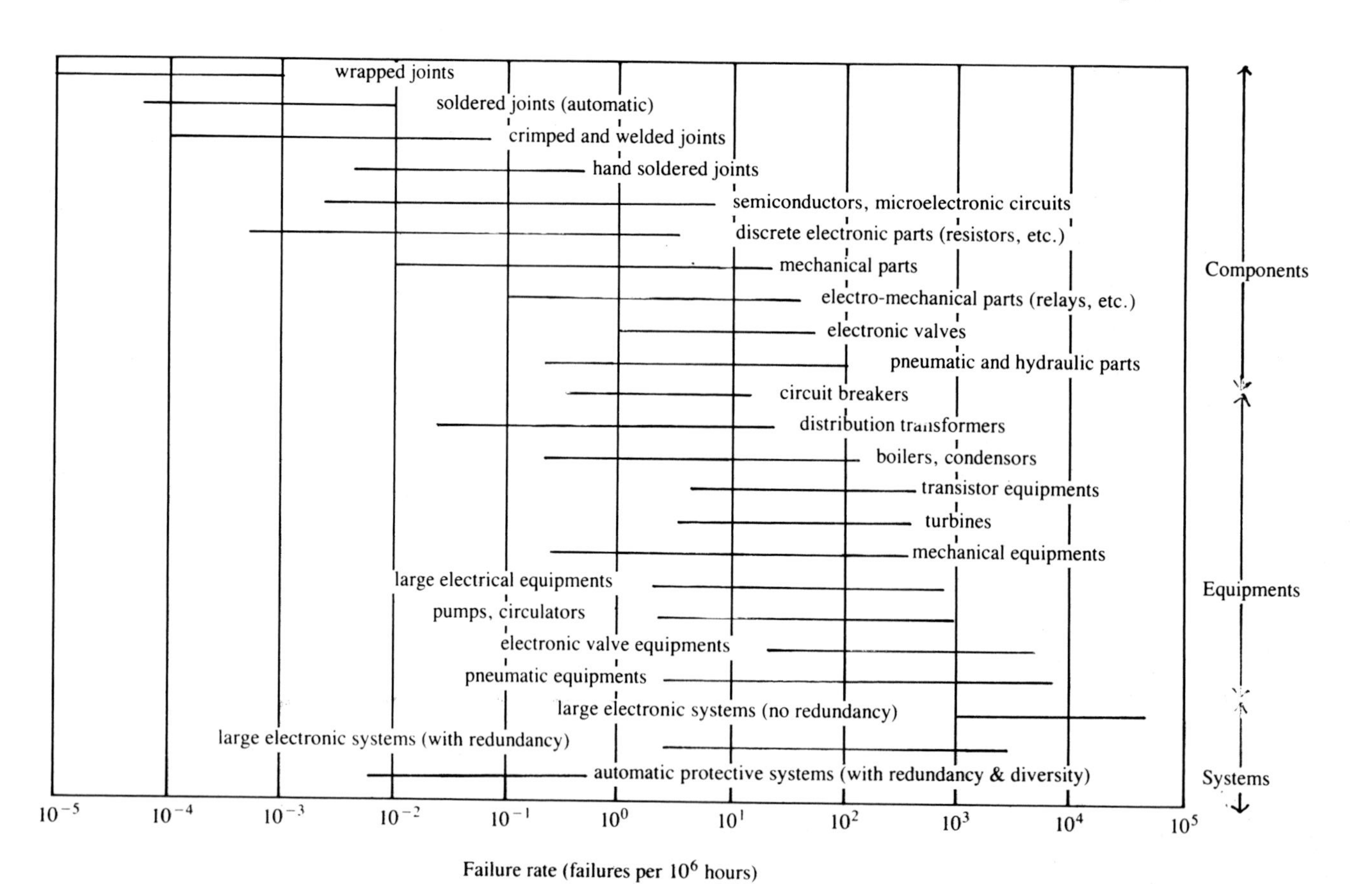

Fig. 19.1. Failure rate changes

necessary to define four specific modes of failure – each associated with a significantly different Mean Time To Restore (MTTR) – in order to generate a realistic representation of system availability.

Information on hazardous incidents has also been shown to be necessary to support a wide range of safety studies. The tendency now is, therefore, toward the extension of the reliability data bank concept to a comprehensive information system which can be used to ensure consistency in the data used in reliability assessments. HARIS (Hazards and Reliability Information System) (**30**), which comprises four data bases for Reliability, Maintainability, Hazards, and Report Abstracts, is one such system.

CHAPTER 20. RELIABILITY DATA COLLECTION SYSTEMS

20.1 General

In order to be able to produce the reliability data necessary to use quantitative techniques, certain basic information is required. This information can best be described as taking three main forms – inventory, operational, and event.

In most large organizations, with fairly large inventories of mechanical equipment, the majority of this information will be recorded by existing schemes, such as log books, fault reports work cards, and the like.

In addition, operators of potentially hazardous plant are required by law to have robust relevant procedures for ensuring the safe operation of their plant. As part of these procedures it is usual to keep operational and maintenance records.

However, these recording systems are generally not suitable for producing quantitative reliability data. Frequently the reporting requirements have grown over the years in response to other management information needs. This is usually achieved, not be revising, reviewing, or rationalizing, but rather by extending the reporting by adding extra tiers of paperwork. This can result in repetition of the reported information and, perhaps, a lack of detail in the relevant areas.

20.2 Inventory data

This data consists of a set of information identifying each piece of equipment by some means of coding, and a record of its design, construction, operation, and environment.

As most mechanical equipment is subject to routine maintenance, the above details should be readily available. The exact form of the record will vary, but should include the following basic information: type of equipment, where it is installed, how it was designed, manufactured, installed, and operated. Essentially this can be achieved by considering the following parameters.

(1) Identification parameter. For example:
   tag number;
   unique identification number;
   generic class;
   location;
   function.

(2) Manufacturing and design parameters. For example:
   manufacturer;
   model/size;
   date of manufacture;
   date of installation;
   design code;
   installation code.

(3) Maintenance and test parameters. For example:
   maintenance schedule and frequency;
   test schedule and frequency.

(4) Engineering and process parameters. For example:
   materials;
   components;
   speed;
   pressure;
   flow;
   temperature;
   process fluid.

The first three parameters constitute a standard set of information which will be similar for all types of equipment. The engineering and process parameters will be defined by referring to each specific class of equipment. For example, for the item class 'centrifugal pumps' the latter parameter set could be as follows.

Engineering:	Body material	Process:	Flow rate
	Impeller material		Suction pressure
	Seal type		Discharge pressure
	Bearing type		Temperature
	Lubrication type		NPSH
	Number of stages		Load factor
	Impeller type		Media
	Coupling type		
	Rotating speed		
	Driver type		

The full inventory data may be stored either in full or in reduced form with indexes referring to other files. The separate files can either be manual or computer based with support from drawings where necessary.

20.3 Operational data

This set of information records the history of the operation of the equipment, and is usually time based (alternatives are cyclic or mission based records). The information will allow the times at risk and lifetimes for each equipment to be determined and should include the following parameters:

Date and times	Installation	
	Operating mode	
	Unavailable due to	– Maintenance
		– Test
		– Failure
		– Modification
		– Replacement
	Standby mode	
	Cycle numbers	
	Mission numbers	

20.4 Event data

This set of information records events occurring on the items contained in the inventory data. These events will normally change the operational state of the item, whether it is due to failure, modification or replacement. The information required is as follows.

(a)	Failure	Mode Cause Consequences How discovered Trade Manhours Type
(b)	Modification	Reason Trade Manhours
(c)	Replacement	Reason Trade Manhours

In order to be able to relate the events to specific items in the inventory file the item identification will be used. Therefore, this data set will also contain the necessary data in a relevant form.

20.5 Reliability data from maintenance records

If the time and effort available for the collection of data is limited, then recourse must be made to the information already available. For most types of mechanical equipment, a record of the maintenance carried out is kept, and this can be used to provide some reliability data. Traditionally the information was kept in a card index or equipment record sheet format, but in most large organizations the records are now stored on a computer system.

The major problem when extracting reliability data in this way is that the information required is dispersed through a wide variety of files. These may differ not only in the recording method (manual or computer), but also in the format, detail, and consistency. This will, therefore, cause difficulty in consistent interpretation.

Population statistics are particularly difficult to obtain – for example, small valves – and, in some cases, estimation of the number of items from the drawings available is the only option.

Operating time is also difficult to obtain as this is usually recorded in many different ways. For example, with rotating machinery a log is normally kept of start and stop dates and times, but in diary format. The hours run will also be recorded using the counter readings provided for this purpose. A final source is the routine statistical returns usually completed for financial purposes.

Failure data is normally extracted from fault reports or work cards. However, with data collection based on maintenance records, the cause and effect of failure are not usually recorded, but the repair action is. The results of scheduled maintenance or routine tests are not rigorously recorded, and any failure experienced during these activities is not always included in maintenance data schemes. Repair times or restoration times are usually available only in the form of manhours, which could give an inaccurate picture. Waiting time is not usually included.

The major advantage in collecting reliability data in this way is that the existing recording schemes are used, and no additional paperwork is necessary. However, a large amount of manual effort is required to coordinate the various sources of data and avoid any omissions.

Fairly recent developments frequently involve the use of a microcomputer and database management system as an add-on facility to existing computerized maintenance schemes. This approach can be adopted regardless of the original source of the data, and greatly simplifies both collection and processing of data.

20.6 Formal reliability data collection schemes

The ideal method of collecting data of any kind is to have a scheme designed specifically for the purpose. In this way the boundaries, conditions, and limitations of the data are all known.

The information required for a reliability data collection scheme was detailed in sections 20.2–20.4

In setting up a formal scheme, the existing systems of recording should be studied to identify what information is already available. The revised data reporting requirements can then be established and a compromise made between what is required and what can be reasonably achieved.

The scheme will usually rely on paperwork procedures of some description, particularly if integration with a maintenance recording system is required. When designing the paperwork, consideration needs to be given to the objectives of the collection system and the level of detail necessary to achieve these objectives.

It is important to involve all personnel who will deal with the proposed scheme in the design of the system, particularly the operations and maintenance staff, as ideally they should complete the major part of the reporting forms. It is unrealistic to expect clerical or even technical staff who are remote from the plant to be primarily responsible for providing the data. Any checking or coding of information is best performed as a centralised function, as this removes some of the effort from operational staff and also ensures a consistency in approach. The storage and retrieval functions should also be carried out in the same manner, although access to the data should be freely available. It is essential that information is fed rapidly to all levels of staff requiring the output, including those who complete the paperwork. They should, therefore, see the benefits of the scheme.

20.7 Input of data

The usual method of setting up and maintaining the inventory, operational, and event data files is to use a data collection form. This can then be used as the input document if a computer based scheme is employed.

For ease of completion of the above paperwork the

majority of the form should be coded information or include tick-box type questions. It is important to strike a balance between the number of choices of codes and the detail required in order to simplify the task of completing the form. The definition of each code should also be chosen with care in order to avoid overlapping or conflicting definitions. It is very easy for certain categories to become frequently used because they are catch-alls. An example of this is the 'how discovered' category; both 'by operator' and 'by routine visual inspection' cover some common ground.

The form should include a free-speech section which, even if not input into the computer, would allow the maintenance staff to add any relevant comments, notes on settings, parameters, or recommendations for improvement. This section often provides useful background information in the analysis of a problem area. Additional cross referencing to other documents is also beneficial, for example continuation forms, other reports, stores requisitions, and so on.

A typical event data form which could be used in many applications is shown in Fig. 20.1

However the use of two forms should not be discounted, as this often simplifies the analysis procedure. This is particularly relevant to mechanical equipment, where it is usual to keep the operating log (recording utilization and availability) and the event data records separate. This ensures that the time or event data-base is adequately recorded.

An example of the paperwork for a formal data collection scheme to provide records on gas compression plant operation is shown in Figs 20.2, 20.3 and 20.4.

20.8 Data retrieval and processing

The outputs provided by a reliability data collection scheme can vary greatly from qualitative to fully quantitative. The exact form will depend on the intended user; the reliability practitioner will normally require the production of reliability parameters (quantitative), the maintenance engineer may only require an indication of problem areas (qualitative).

In order to derive the reliability parameters (that is, failure rates, failure modes, repair rates, and so on) for selected samples of relevant components, access is required to information in the three major files, that is, Inventory Data file, Intermediate file, and Event Report file. The link between these files would be the identification data which should be common to all records in different files.

The selection of the relevant records from the Inventory Data file should be possible at the desired level of detail, the two extreme levels being a unique item selection (by means of tag or unique identification number) or an overall component class selection (by means of the generic class code). Intermediate levels are those specifying the generic class code (for example, Centrifugal pumps) plus one or more parameters of the inventory data sheet (that is, manufacturer, media, rotational speed, and on on).

The most useful tool for making such a selection will be a computer program capable of searching the Inventory Data file by the parameters specified. Once the program has identified the relevant inventory sheets, their content together with the associated Event Reports should then be transferred into an Intermediate file for further processing. The Event Reports associated with the selected sample of items are identifiable via their tag or unique identification number or generic class code. Depending on the purpose of the analysis either all the Event Reports will be transferred into the intermediate file, or only those having pre-defined parameters, that is, those dealing with a specified failure mode. Thus, the program should be capable of searching the Event Report file at the desired level of detail for events associated with the selected inventory items. The content of the Intermediate file will then be processed manually, or by suitable statistical analysis programs, and the relevant parameters derived.

This data retrieval and processing system must be flexible, having the capability of producing either generic data (for example, failure rate of centrifugal pumps) or very detailed data (for example, failure rate of centrifugal pumps manufactured by [say] Worthington, on seawater service, with rotating speed up to 3000 r/min, when the failure mode was major leakage from the seals).

In large organizations the data files and processing will all be based on mainframe computers. For small organizations, the use of a microcomputer and a commercial data-base program will provide the necessary facilities.

20.9 Checklist of procedures for setting up a data collection scheme

(a) Identify the generic classes of items on which reliability data is required.

(b) Define the physical boundaries of each equipment class.

(c) Compile lists of tag/unique identification numbers to establish populations for each generic class.

(d) Define minimum sample sizes and the reliability parameters to be derived.

(e) List the event data input required to derive the required reliability parameter output.

(f) List the assumptions to be made in analysing the event data and the tests proposed to validate these assumptions.

(g) Define the terms and develop coding to be used on the data collection forms.

(h) Develop information flow diagrams tracing the routes from input of the basic data to reliability parameter output.

(i) Develop procedures for collecting and inputting data into the system.

(j) Inform and involve all staff affected by the scheme and in its implementation.

(k) Carry out a pilot exercise to identify any problem areas and modify the procedures, if necessary, and start data collection.

EVENT REPORT FORM	
ITEM IDENTIFICATION DATA Tag. No. Unit ID No. Generic code	REPORT No. COMPLETED BY: APPROVED BY: DATE:
EVENT TYPE:	
TIME ALLOCATION DATE TIME FAILURE DETECTION: START MAINT. ACTION: COMPLETE MAINT. ACTION: READY FOR OPERATION:	
FAILURE MODE 1. 2. 3. 4. 5.	
EFFECT ON SYSTEM 1. 2. 3.	
RESTORATION MODE 1. 2. 3. 4. 5.	ENGINEERING CRAFT HOURS 1. MECHANICAL 2. ELECTRICAL 3. INSTRUMENTS 4. OTHERS
ENVIRONMENT High Normal Low Ambient temperature Humidity Dust Vibration	
EVENT DESCRIPTION (text)	

(Note other important environmental factors which could contribute to failure in Event Description)

Fig. 20.1. Event report form

OPERATIONS DEPT. UNIT OPERATIONAL LOG SHEET.

PERIOD FROM 0600 ON [DAY | MTH | YEAR] TO 0600 ON [DAY | MTH | YEAR] STATION [ ] UNIT [ ]

	START				STOP							MAINTENANCE										
								RUNNING HOUR METERS (TO NEAREST HOUR)						START JOB			COMPLETE JOB					
	DAY	MONTH	TIME	REASON	DAY	MONTH	TIME	TOTAL	LOADED	PEAK LOADED	REASON	JOB CARD NUMBER	STATE	DAY	MONTH	TIME	DAY	MONTH	TIME	TRIP CODE	UNIT VENT	REMARKS
B/FWD.																						

REASON FOR START	REASON FOR STOP	PLANT STATE	UNIT VENTS
R = ROUTINE TEST M = MAINTENANCE TEST S = SPECIAL TEST G = GRID O = OPERATOR TRAINING	N = NORMAL STOP M = PLANNED MAINTENANCE STOP S = STARTING TRIP R = RUNNING TRIP	S = STANDBY O = OUT OF COMMISSION	EACH TIME UNIT IS VENTED. ENTER V IN VENT COLUMN.

FILL IN THIS SECTION ONLY WHEN A MAJOR COMPONENT IS CHANGED.
G = GAS GENERATOR
P = POWER TURBINE
C = GAS COMPRESSOR

CODE	CHANGED AT UNIT TOTAL HOURS	DETAILS OF REPLACEMENT PREVIOUS TOTAL HOURS	DETAILS OF REPLACEMENT SERIAL NUMBER

Fig. 20.2. Operational data report form

SERIAL NUMBER	LOCATION	DATE RAISED DAY	MONTH	YEAR	UNIT CODE	TRADE	HOW DISCOVERED	SERIAL NUMBER OF CONTINUATION SHEET

SYMPTOMS AND / OR TRIP INDICATION

RAISED BY ....

RECTIFICATION

DATE COMPLETED ____________ MAN HOURS TO COMPLETE __________ HOURS RUN ____________ SIGNATURE ....

MANUFACTURERS NAME ______________________________

COMPONENT DESCRIPTION ______________________________
(INCLUDING SIZE, TYPE NO WHERE RELEVANT.) ______________________ NO. OF COMPONENTS USED ....

Fig. 20.3. Event report form (tradesman's copy)

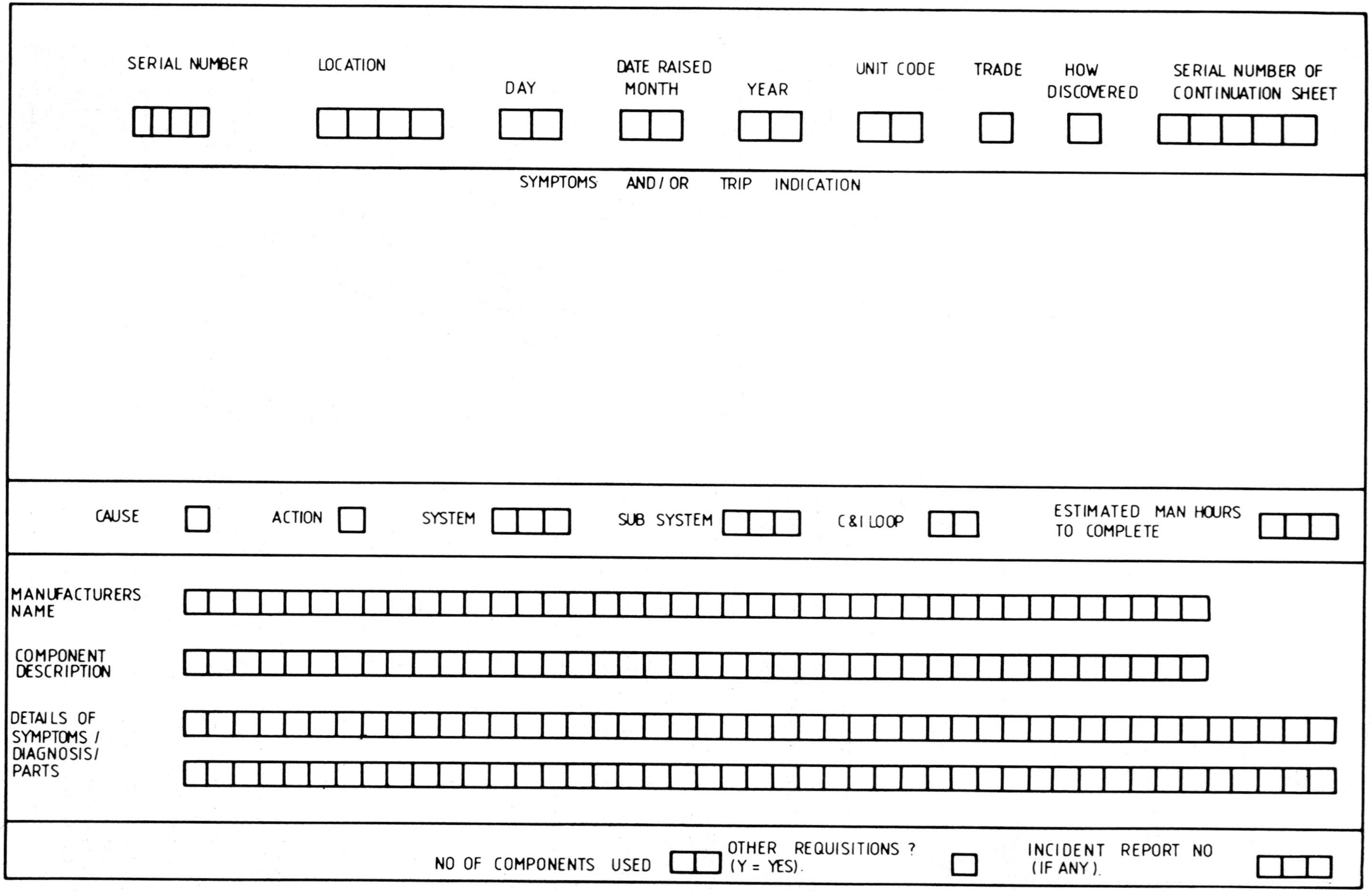

SERIAL NUMBER

LOCATION

DATE RAISED: DAY MONTH YEAR

UNIT CODE

TRADE

HOW DISCOVERED

SERIAL NUMBER OF CONTINUATION SHEET

SYMPTOMS AND/OR TRIP INDICATION

CAUSE

ACTION

SYSTEM

SUB SYSTEM

C&I LOOP

ESTIMATED MAN HOURS TO COMPLETE

MANUFACTURERS NAME

COMPONENT DESCRIPTION

DETAILS OF SYMPTOMS / DIAGNOSIS/ PARTS

NO OF COMPONENTS USED

OTHER REQUISITIONS ? (Y = YES).

INCIDENT REPORT NO (IF ANY).

Fig. 20.4. Event report form (input copy)

20.10 Some further considerations

(a) Feedback and circulation

A drawback of any form of data collection scheme, is that the staff who complete the paperwork, can usually see no immediate benefits. It is important to integrate any necessary paperwork with other schemes, which, hopefully, will reduce the numbers of forms to be completed. It will also be beneficial to give adequate training before implementing the scheme, and at this stage the future benefits can be explained. When information from the scheme reaches a useful level, it is important that this is circulated promptly to all personnel involved.

Data collection and processing is an important activity requiring dedication and intelligence. It needs to be implemented and operated by experienced engineers who understand the objectives, and the benefits, to be derived from the collection and analysis of reliability data. This must be recognised by senior management and the appropriate level of staff allocated to the project. Without this commitment, it is unlikely to be cost-effective and will soon wither and die.

The analysis of the data should preferably be carried out by specialist staff, again to save effort on the part of operational personnel. A central analysis department will also aid communication between differing locations.

(b) Limitations and assumptions

The majority of data collected represents a period within the lifetime of a plant, that is, the plant had already been in operation for some time before data collection commenced and is, presumably, continuing to operate. It is therefore doubly censored, and this effect is covered in Part Two.

Because of the above, and the difficulty in obtaining lifetime distributions for each component, it is generally assumed that failures are occurring randomly with respect to time. This implies a Poisson process and the arithmetic mean gives the best estimate of failure rate.

(c) Costs and benefits

The main argument against collecting reliability data is one of cost. The information already collected on mechanical equipment is probably not designed to provide reliability data and, therefore, if required, it will be a costly exercise to acquire it.

Plant operators and managers, quite justifiably, need to be able to see that the financial benefits of providing and using reliability data outweigh the costs involved in obtaining it. The cost of collecting the data will depend very much on the size and scope of the operation, but as an indication the following could be taken as a guide:

(i)	Microcomputer and peripherals	£5000
(ii)	Database software (for approx. 2000 records)	£ 500 (at 1986 prices)
(iii)	Manpower 1 clerical person full time for analysis and data input.	

It should be noted that, provided that the paperwork procedures are properly designed, taking full account of other requirements such as maintenance planning, the additional effort required on the part of maintenance personnel or operators should not be significant.

These data collection costs can be offset against the improvement in reliability and availability seen as a result of collecting data and applying the results of analyses and studies. The effect of this should be to:

reduce maintenance activity;

reduce the standby equipment required;

minimize lost revenue;

improve safety levels or maintain existing safety levels with less equipment.

PART SIX

Case studies

INTRODUCTION

The case studies included have been culled from the experience of the contributing authors in dealing with reliability and availability problems within their own, or from client, organizations. The purpose of including the case studies is to enhance the reader's understanding of the theory and techniques, which are covered in the main text, applied to real industrial problems.

The particular case studies presented have been chosen because they illustrate certain commonplace events which occur in the development of custom designed and built process plants. Some of these are listed below.

(1) When major new capital expenditure is proposed the owner/operator wishes to know what the likely level of plant reliability and availability will be, which major items of plant are likely to contribute most to plant downtime, and whether there is an economical case for increasing reliability and availability, either through redesign, using redundant plant, or improving the levels of spares.

(2) During the early stages of commissioning and debugging of new plant, as priority problems are resolved, other more long-term reliability problems emerge, where each individual event has a relatively small impact.

(3) Particularly in the case of continuously operated process plant, when plant extensions are proposed, one of the first questions asked is 'How will this effect the availability of the main plant?'

They have also been chosen because they illustrate the main themes of this publication.

(1) That reliability engineering is a predictive, probabilistic, applied science, and that the collection and analysis of past operating experience is used to predict, and in some cases shape, future events.

(2) That the *application* of reliability engineering techniques is not as complicated and complex as the background mathematical theory might suggest.

Clearly, it is not possible to give a fully detailed account of the case studies here because of the mass of information involved. The information has been greatly summarized, with examples of analysis on particular parts or items of plant to illustrate the techniques used, and the main emphasis is on the methodology and procedures used.

CASE STUDY No. 1. FAILURE INVESTIGATIONS OF BUS GEARBOXES

A. Kelly *Manchester University*

A local transport company operated a fleet of buses, 42 of which used a particular type of gearbox. It was the policy of the company to extend the use of this particular gearbox across the whole of its fleet since it was designed to the company's specifications.

An investigation of the vehicle history files and discussions with the garage maintenance staff revealed that the gearbox, both new and overhauled, was one of the high-cost areas in corrective maintenance.

The make of vehicle using this gearbox had been introduced in mid-1975 and, since only a small number of gearboxes had proceeded beyond the second overhaul, it was decided to classify them into two categories only, viz: (i) new and (ii) overhauled, which latter included all gearboxes that had been overhauled once or more.

1 Gearbox description

Figure CS1.1 shows the gearbox transmitting power in 1st gear position. The gear trains are of basic epicycle design. Each gear position is engaged by clamping a brake around the relevant annulus to hold the annulus stationary. The selection of the gear is via the electro-pneumatic selection mechanism shown in Fig. CS1.2. The gearbox employs the fully-charged fluid-coupling principle, whereby oil is common to the fluid coupling, gearbox, and angle drive. Thus, five brake bands are used, including that for reverse gear. To allow for brake band liner wear, an automatic adjustment mechanism is incorporated into the operating linkage. A multi-plate clutch is used for the fifth gear, consisting of phosphor bronze friction plates mating onto steel plates.

2 Data collection methods

The existing data collection system recorded a four-weekly mileage and cumulative mileage for each vehicle, gearbox, and other major items. Unfortunately, mileage to failure was not recorded. A good estimation could be obtained, however, in the case of a gearbox, from the cumulative mileage and the failure data.

Gearbox failure data, cause of failure, and components replaced during each overhaul were supposed to be recorded on an 'item change-over-card' and vehicle history file. However, the quality of the data collected was poor and, in some cases, non-existent. Little attention was given to data analysis. As a result of this, unnecessary time was spent extracting the failure mileage from the various files.

Since the causes of failure were not clearly recorded, it was necessary to observe the failed gearboxes being overhauled and to carry out detailed discussions with the workforce involved in order to understand the mechanisms, and, where possible, the causes of failure.

3 Failure analysis

3.1 Gearbox failures

The on-site work carried out by the author revealed that the gearboxes (new and reconditioned) could fail in one of six ways (see Figs CS1.3 and CS1.4).

(1) Brake band fracture – due to: (a) the band being over-stressed because of incorrect band adjustment, inadequate pressure, malfunction, and violent gearchange, (b) fatigue as the clamping and releasing of the band gave rise to alternating stresses, (c) stress concentration at sudden changes in section, sharp corners, and rivet holes (these could also initiate fatigue cracks), (d) material defects.

(2) Brake band liner fracture – due to fatigue and stress concentration around the rivet holes.

(3) Brake band liner wear – which includes three-body abrasion.

(4) Top gear clutch plates wear and seizure.

(5) Gear train failures – due to pitting caused by rolling and sliding actions.

(6) Other mechanical failures and material defects – such as broken shaft and casing, pump failure, and so on.

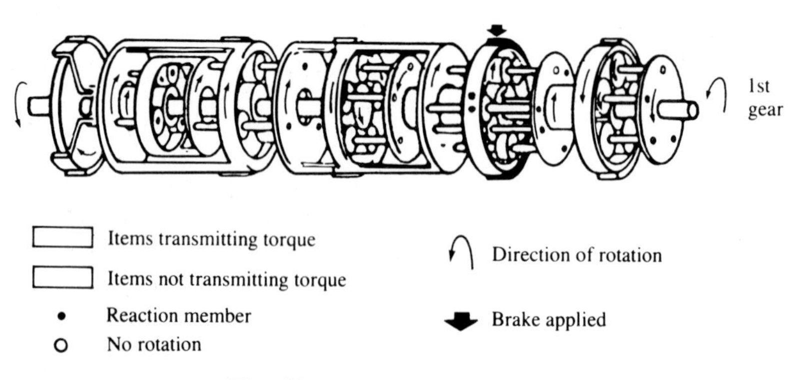

Fig. CS1.1. Power flow diagram

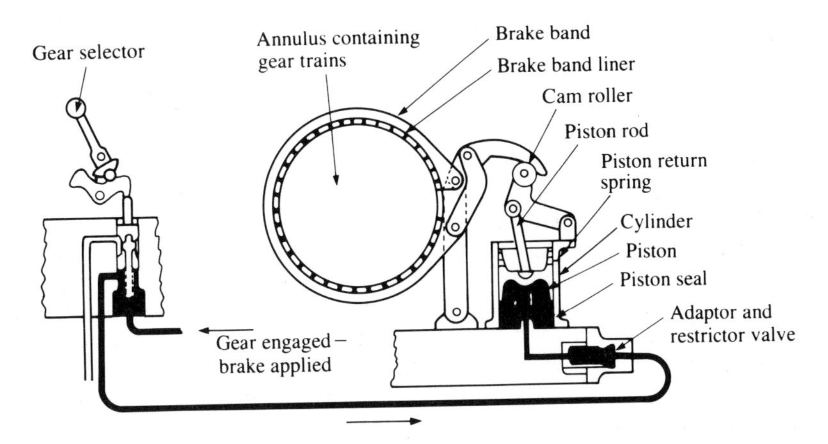

Fig. CS1.2. The air control diagram

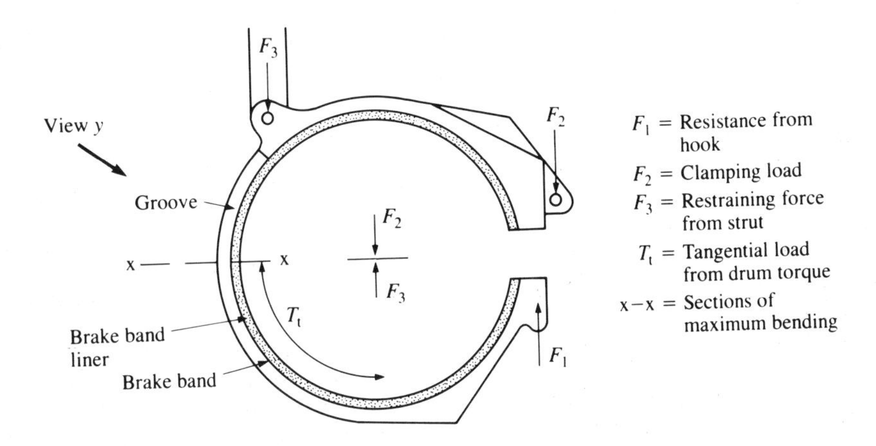

Fig. CS1.3. Schematic diagram of the brake band showing forces applied

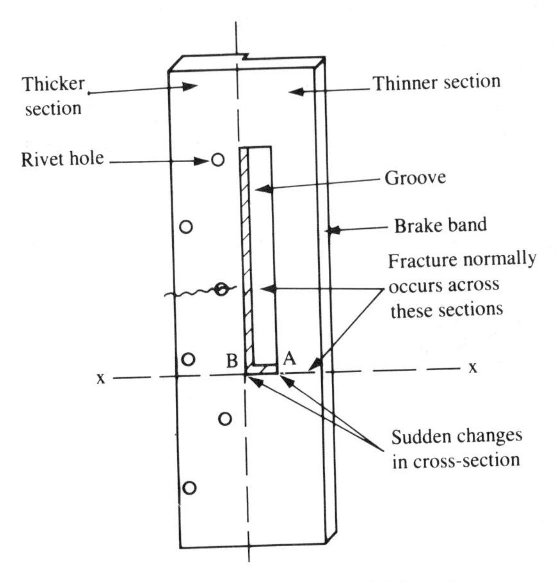

Fig. CS1.4. Brake band (view y)

3.2 Failure mode and causes of failure of new gearboxes

Weibull analysis of the new-gearbox failure data (Fig. CS1.5) gave a shape factor $\beta = 2.25$, indicating wear-out failure. The mean time to failure (MTTF) was 120 000 miles, or approximately 3.1 years service.

The vehicle history records and observations on new gearboxes showed that the *main* causes of failure were brake band wear and top gear clutch plate wear. Brake band liner fracture was also observed, but this did not cause any gearbox failures. Brake band fracture was not observed. These observations, indicating that wear was the dominant failure mode, confirmed the validity of the observed Weibull shape factor.

The MTTF, 120 000 miles, was less than the manufacturer's claimed life of 200 000 miles. This low value was attributed to the following factors.

(1) The lubricant had poor oxidation resistance, no dispersant or detergent properties, unsatisfactory anti-wear and extreme pressure properties, and inadequate low temperature fluidity. This resulted

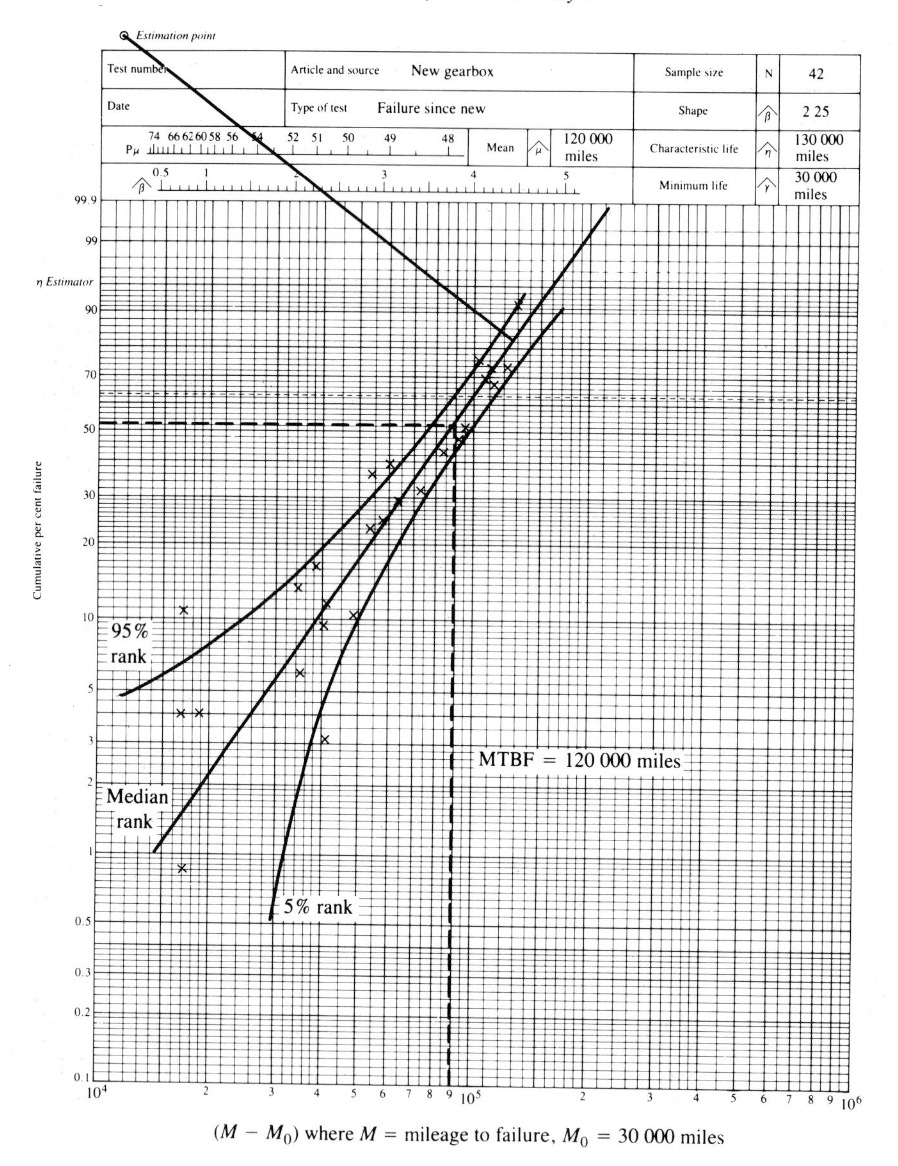

Fig. CS1.5. Weibull plot for new gearbox failure

in sludge formation, accelerating three-body abrasion.

(2) The period of lubricant replacement had been set at 24 weeks, or 19 000 miles (approximately) and was too long. The manufacturer had recommended that the gearbox oil should be changed at 12 500 miles.

(3) Poor maintenance of the air control system resulted in low air pressure in the transmission system (inadequate checks) and incorrect adjustment of the piston travel. This caused brake band slip and excessive wear of the liners.

(4) Driver misuse, such as over-revving during rapid downward gear change. Full throttle upward gear change and idling in-gear also caused excessive wear. Idling in-gear was particularly detrimental to top gear clutch plates, causing excessive wear and heat dissipation.

The solutions recommended to the operator were as follows.

(1) Implementation of better and more effective maintenance procedures, including the following.
   (i) More frequent gearbox oil change, reducing the 24-week usage period to 18 weeks. The correct periodicity could be determined by lubricant analysis. The long-term solution would be to find a more suitable gearbox oil.
   (ii) Checking and adjustment of all piston travel and brake band automatic adjusters every 18 weeks, when changing the gearbox oil.
   (iii) Checking and rectifying air leaks in the trans-

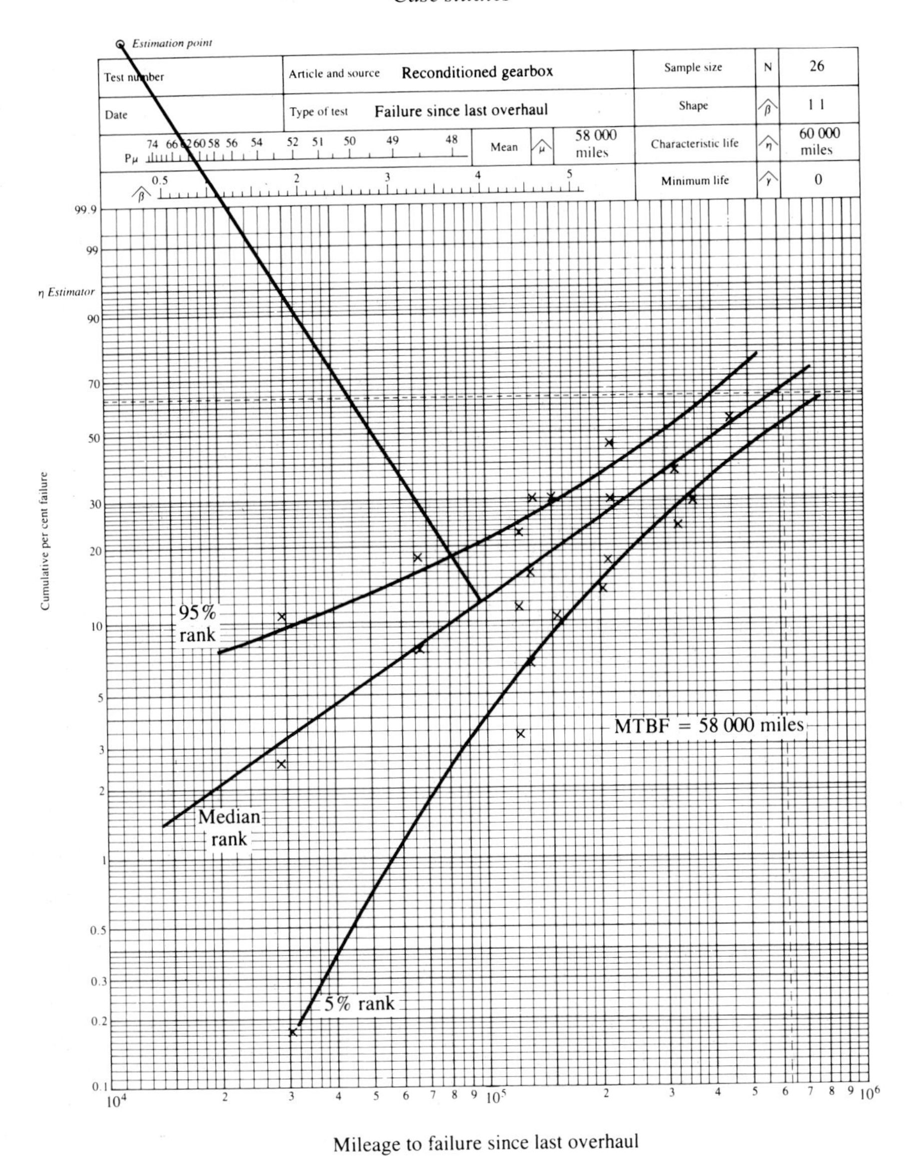

Fig. CS1.6. Weibull plot for reconditioned gearbox failure

mission system every 6 weeks to ensure correct air pressure. Air leaks should be checked objectively (for example, using a shock pulse meter with special probes so that even minor leaks would be detected) instead of the existing subjective, aural method.

(2) Training and education of drivers in the correct handling of the vehicle.

(3) Designing brake band liners that would be thicker and more wear resistant. It has been shown that the life of another make of gearbox which used a thicker brake band liner (with the same frictional area and material) was double that of the gearbox studied.

(4) Reducing stress concentration by eliminating sudden changes in cross-section and sharp corners on brake bands (see Fig. CS1.4).

3.3 Failure mode and causes of failure of reconditioned gearboxes

Weibull analysis of the failure data for the reconditioned gearboxes (Fig. CS1.6) gave a shape factor $\beta = 1.1$, indicating random failure. The MTTF from reconditioning was 58 000 miles, or approximately 1.5 years service. The life of a reconditioned gearbox was, therefore, only half that of a new one.

The main causes of failure were brake band fractures (especially on the second gear) and top gear clutch plate seizure, the former being much more frequent. Pitting-fatigue of the gears, especially on the second gear, was observed, but no tooth fracture occurred. There were no failures due to brake-band liner or top-gear clutch-plate wear.

An engineering investigation showed that fracture occurred in the inner band section and across one of the rivet holes in the thicker outer band section. It was concluded that, due to alternating loads, fatigue cracks (initiated by the stress concentration in the inner band section) propagated slowly until the inner band could no longer withstand the load, and tore off. At this point the thicker section fractured across the rivet holes.

The low life and random failure of reconditioned gearboxes resulted mainly from the practice of using reconditioned brake bands when reconditioning the gearbox. A brake band fitted into the 2nd gear position (greatest use and stress) in a reconditioned box might have already operated for a considerable amount of its fatigue life. The remaining life obtained from such bands will vary randomly, being dependent quite fortuitously on previous gear position and previous operating life.

The actions recommended for increasing the MTTF were as follows.

(1) Replacement of all five brake bands with new ones during reconditioning. This would minimize fatigue fracture, but would increase the reconditioning cost by about £500. This latter would, however, result in a considerable potential saving in labour and in bus unavailability costs.

(2) Instrumented crack-detection surveys of all gear trains and shafts before refitting. The previously used visual check was too subjective and would not detect hairline cracks.

(3) Gearbox manufacturer to be asked to reduce stress concentrations by rounding off all sharp edges on the brake bands.

4. Discussion

The experience gained from this study showed that, in order to control the level of plant failures, an organization needs to adopt a systematic procedure for the recording, storing, analysing, and feedback of failure data. Such a procedure involves three main levels of control: the shop floor and first-line supervision, the history record and spares system, and the equipment manufacturer.

The advantages to be gained at the first level of control were shown in this study. Considerable data and information were obtained by observing repairs and reconditioning procedures and by discussing the history of failures with tradesmen and supervisors.

The study also showed that, unless a data collection system has been properly designed, it is extremely difficult to extract the type of information necessary for maintenance decision making. A passenger transport organization using many identical buses should have a data collection system which will gather information on failures down to the item (for example, gearbox), level of plant and, where necessary, down to component level. Such information should include the times-to-failure, the symptoms, and, above all, the causes of failure (where known). Through the use of computers it is now a straightforward matter to arrange for indication, on a monthly basis (or less) of the main problem areas, in terms of frequency of failures or cost. The spares system can also be designed to be used as an additional source of information. For example, it can indicate high spare-part usage-rate. In each case, determination of the cause of failure should be followed by selection of the most appropriate corrective action. Where this has been achieved, the information should be passed to the manufacturer, either directly or via specifications for new equipment. In the same way, the manufacturer passes such information through to the equipment user.

CASE STUDY No. 2. IMPLEMENTATION OF A RELIABILITY PROGRAMME ON A NEW GAS COMPRESSION STATION

T. P. Littlejohns M. Kelly British Gas Corporation

1. Introduction

The British Gas Corporation operates and controls a complex gas transmission network designed to ensure security of gas supply, in a safe, reliable and efficient manner to meet widely varying summer and winter demand. The network is an integrated, countrywide, system of pipelines, which receives natural gas at shore terminals and transports it at high pressure to regional offtakes for distribution at a low pressure to the customer. At a number of carefully selected points along the high pressure system, a series of aero derivative gas turbine powered compressor units have been installed to increase the flow capacity of the pipeline. At present there are 15 compressor stations, comprising 44 operational machinery units.

Each machinery unit is installed in its own building and consists of a train of three sub-units, namely, a gas generator, a free power turbine, and a centrifugal compressor, as shown in Fig. CS2.1.

The sub-units, with their support systems are mounted on to one or more bedplates connected to form a single substantial structure. The main supporting systems required to operate and monitor the machinery include fuel gas, lubrication, and sealing oil equipment, as well as instrumentation for control and protection. High velocity ventilation ensures a safe environment within the building and the air intake and gas exhaust systems are acoustically designed to maintain acceptable external noise levels.

When selecting machinery we seek to ensure that in operation the equipment will achieve the required performance safely, economically, and reliably. We use standard reliability engineering techniques to quantify reliability in terms of start probability and mean running time between failures (MTBF).

2 Reliability assessment of new plant

Within British Gas it is now common practice to include reliability assessments in plant evaluations. The two most frequently used methods for assessing reliability at the specification and design stage are the Generic Part Count and the Fault Tree Analysis.

(a) *Generic part count technique*
The Part Count method predicts the reliability characteristics of the overall equipment from consideration of its detailed parts. The depth of the

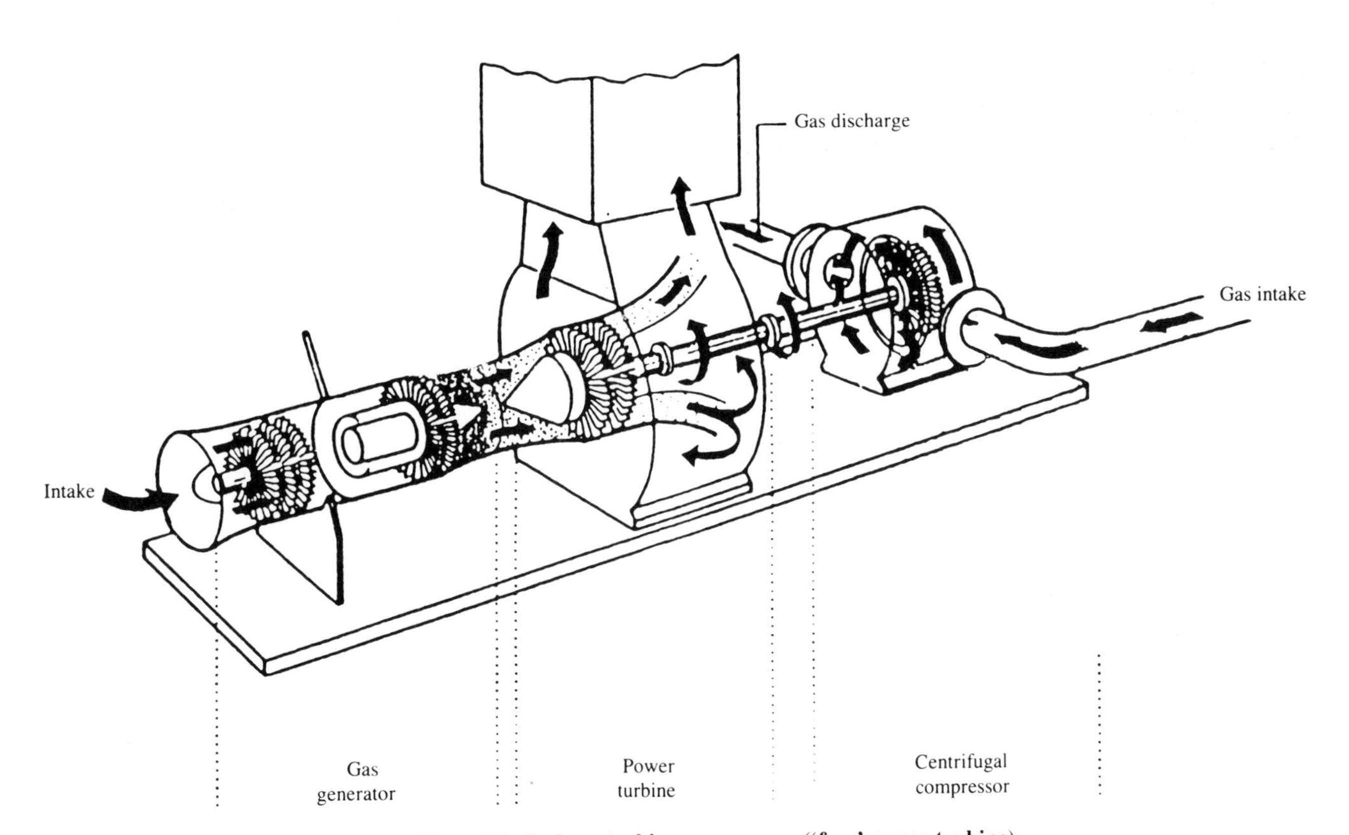

Fig. CS2.1. Typical gas turbine compressor ('free' power turbine)

assessment, is determined by the detail of the reliability information available.

If there is sufficient information available the reliability assessment will consider the failure modes of each part and their effects on the overall equipment performance. This is termed a Failure Modes and Effects Analysis (FMEA).

(b) *Fault tree analysis*
The analysis proceeds by identifying the potential failure situations. The chains of events that lead to these situations are developed, and the reliability of the equipment is predicted from consideration of the probabilities that the combination of events will occur.

The approach chosen will depend on the problem and the data available.

The information required for reliability assessments is generally based on previous experience and obtained from one or more of the following sources.

(a) In-house records.
(b) Cooperation with other plant users who share common experience.
(c) Commercially available data bases. (For example, the United Kingdom Atomic Energy Authority (UKAEA) Systems Reliability Service.)
(d) Published sources.

It is generally possible to obtain a satisfactory reliability assessment in the absence of complete data because the majority of problems arise on a minority of components. Concentrating on these dominant components will usually yield a reasonable assessment.

The assessment is primarily directed towards forecasting the rate of failures that may be expected to arise from operation within the design envelope, due to randomly distributed failures of components. It assumes that:

(a) The equipment is capable of performing its required duty as designed.
(b) There are no latent design problems to consistently cause failure.

The second assumption is critical to the prediction and, as will be shown in this paper, can lead to significant differences between forecast and actual reliability performance.

3 Reliability programme for a new compressor station

3.1 The machinery selection

In general we aim to select well proven equipment. However, for one of our recent two-unit compressor stations, we considered a number of new features and requirements, which called for certain innovations in the type of machinery and supporting equipment selected. The question of reliability was, therefore, brought to the forefront.

The station was the first within British Gas to have the capability of running unmanned, with operation being controlled from a remote centre. The telemetry system needed to support the remote control was fed by a minicomputer based machinery control system. Because of the remote operation, starting reliability was considered to be of paramount importance.

Although the station had a requirement for two machinery units, we proposed to save considerable costs by utilizing an existing refurbished gas generator for one unit. Therefore, it was important to ensure that the reliability of the older machine, when installed and interfaced with new equipment, also met the reliability specification.

The machinery selection was also influenced significantly by the future operating costs. We wanted the maximum available efficiency from the centrifugal compressor to minimize the payback period. A new standard of higher-efficiency machines was available, but improved aerodynamics were often incorporated with other new or updated design features. Reliability assessment of the compressor was, therefore, difficult because little operational experience had been achieved and, hence, the data available was of limited relevance.

3.2 Reliability targets

The following targets were set at the level of the best performance of units then in service, and considered achievable without requiring any additional research and development.

(a) To achieve a start reliability of 0.95 by the end of 200 starts after acceptance.
(b) To achieve a run reliability giving a MTBF of 475 hours by the end of 3000 running hours after acceptance.

To ensure the targets were achieved within the stated operational period, a reliability programme was initiated. The programme was a cooperative venture with the supplier, but one in which the onus for the achievement of the specification rested with the supplier. The programme consisted of three phases.

(a) A design assessment to identify potential problem areas and, if possible, eliminate these prior to manufacture and installation.
(b) Works test to confirm functional operation.
(c) An operational trial to evaluate reliability and identify and correct problems to achieve the specified targets.

The programme was managed by a joint British Gas Corporation/Supplier Reliability Committee. During the operational trial, members of this committee met periodically to review the operational data and allocate, by agreement, failures to the supplier's equipment.

4 Design assessment

The design assessment consisted of:

(a) A reliability prediction.

(b) The identification of problems on similar units already in service.

The reliability prediction employed the part count technique and drew on the most applicable published data for failure rates. It was supported by limited Failure Modes and Effects Analysis (FMEA) in some areas. Of necessity the prediction contained a number of simplifying assumptions. In particular, that failures would occur randomly and that all major components would operate without latent design problems.

The analysis predicted that the start reliability target would be met, that is, the machines would provide a 95 per cent start reliability. The run reliability targets were considered more difficult to achieve. However, no single area was identified as the most significant influence.

Reliability data from units in service at the start of the design assessment was limited. However, as relevant information became available it generally supported the results of the reliability prediction. In addition it highlighted a number of minor potential problem areas and corrective action was implemented.

The reliability prediction and review of in service problems generated the following changes to the proposed design.

(a) Minor modifications to the seal oil control and lube oil control systems.

(b) Changes of components within the fuel gas and power supply systems.

(c) A reduction in the number of shutdown protection systems (that is, trips). A total of 10 trips were deleted or downgraded to alarms.

(d) The installation of a dedicated chart recorder for start failure diagnosis.

In addition there was an increase in the scope and depth of quality assurance checks to ensure the fitness of critical equipment, especially where problems had occurred on earlier units. The oil systems, electronics and instrumentation were the areas where most of the additional effort was applied.

5 Preliminary observations from the works tests

5.1 The gas generator and power turbine

The gas generator and power turbine were successfully tested by the main supplier. Despite rigorous control system checks, the tests gave no cause for concern regarding the validity of the reliability assessment. However, we were still far from operating the machines under true site conditions.

5.2 The compressor

The centrifugal compressor tests were conducted separately from the gas generator and power turbine. Preliminary results from the compressor tests indicated that the machines were not running in a satisfactory manner. Performance efficiency was less than predicted, and mechanical problems were resulting in excessive vibrations.

After a number of visits to the vendor it became obvious that the performance deficiency could be solved, whereas the mechanical problems were fundamental to the design of the machine. Thus we were entering a development investigation of new equipment and the assumptions of the original reliability assessment were no longer valid.

5.3 The compressor problems

The basis of the mechanical problems lay within a combination of shortcomings in the seal oil system design and the compressor rotor dynamics. The seal elements were not working correctly, but their problems were being compounded by a shaft system instability.

The seal system is shown in Fig. CS2.2. The inner carbon face seal is on the high pressure gas side and the two outer floating bushing rings seal to atmosphere. The space between inner and outer seals is filled with high pressure oil, controlled to maintain a positive oil-to-gas pressure across the face seal. The gas pressure can vary from 40 bar up to 70 bar. This arrangement is needed at both ends of the shaft adjacent to each bearing assembly.

Oil pressure control is vital. Excessive pressure will push oil past the face seal and flood the compressor. Too little pressure will allow gas into the seal and bearing assemblies, resulting in a highly volatile oil and gas mixture passing to other parts of the machinery, including the hot turbine.

The floating bushing seals were at the root of the problem. Due to the high differential pressure across each ring they were locking up within the seal housing, but only under certain unknown conditions. The loading forces across one bushing ring is shown in Fig. CS2.3(a). Their lack of ability to float and move to cater for any small shaft movements was having a detrimental effect on the shaft vibration characteristics. It appeared that the rotor shaft system was not always being simply supported by the radial bearings but was being constrained by the two sets of bushing rings. In effect the bushing rings were acting as another set of bearings at each end of the shaft, producing different and detrimental shaft dynamics.

The first action taken was to off-load the high axial thrust on the bushing rings. This was achieved by reprofiling part of the ring and bleeding high pressure oil to the rear face. The resulting pressure profile is shown in Fig. CS2.3(b). The success of the modification was only partial. The anticipated reduction in vibration levels was not sufficient to meet the pass-off limits.

The second action was to remove one of the two bushing rings. The result was dramatic. The high vibration reduced to very low and acceptable levels, but although the remaining seal performed well, the seal oil leakage rate increased more than predicted. This latter aspect gave some cause for concern because we were now using most of the available capacity of the oil supply pumps. Nevertheless, it was considered we had overcome our primary vibration problem and, provided the compressor and pumps exhibited similar characteristics under site conditions, our solution would remain valid.

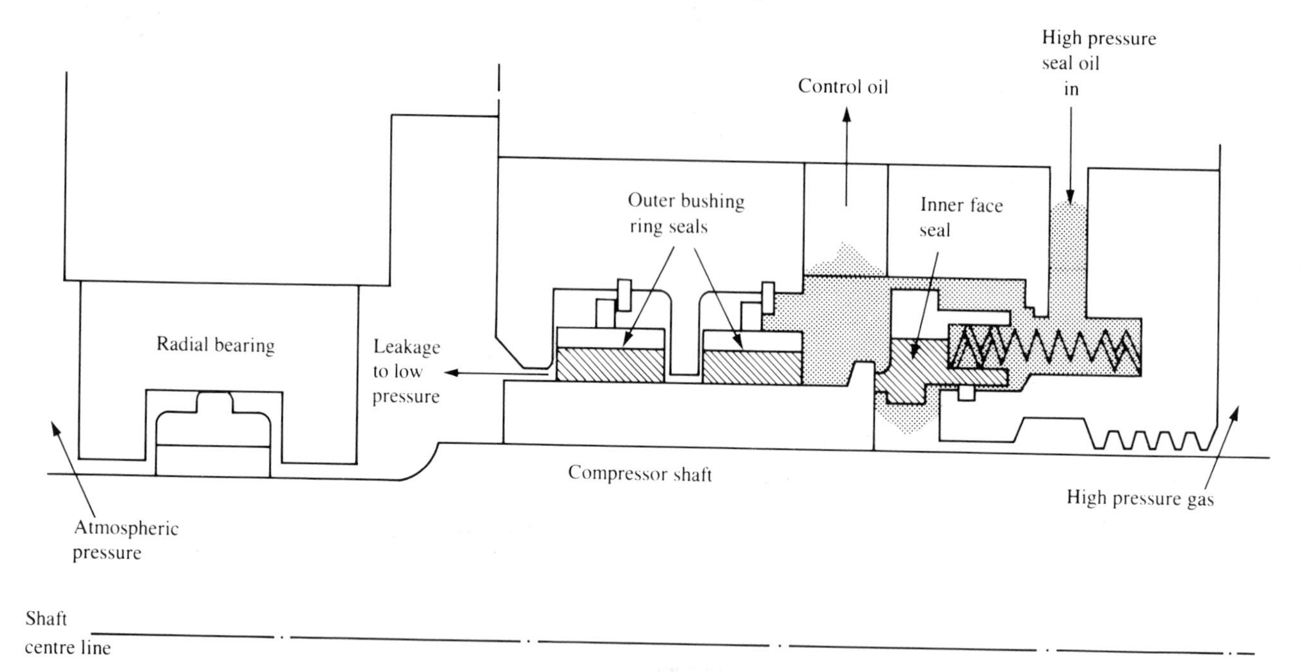

Fig. CS2.2. Compressor seal system and bearing arrangement

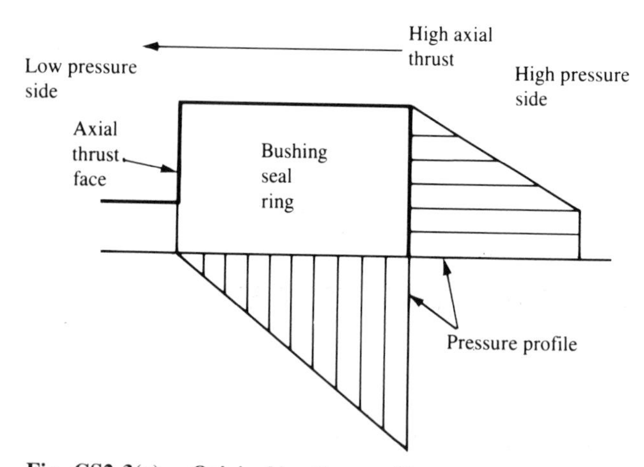

Fig. CS2.3(a). Original loading profile across one seal ring

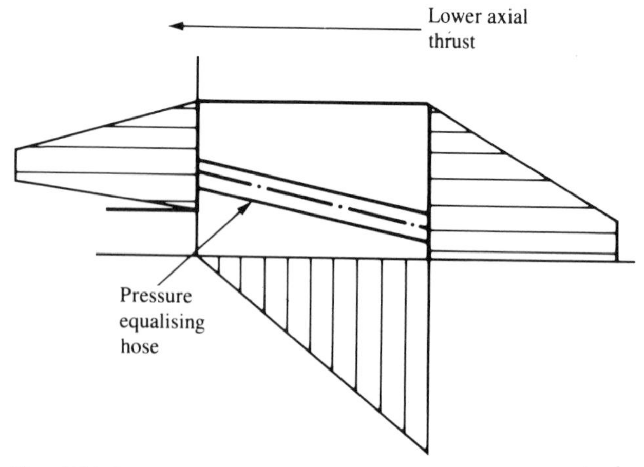

Fig. CS2.3(b). Modified seal ring giving improved loading profile

6 The operational trial

6.1 Reliability data logging

In order to allow for a debugging and 'running-in' process, we had specified that the first 20 starts and 200 hours of operation should be discounted in terms of the reliability trial.

Once the trial had commenced we logged all the unscheduled machine shutdowns or 'trips'. For each trip, a detailed job card was made out, which recorded all relevant information available from the unit control monitoring system. Such information relevant to a start failure would include the start recorder charts, showing fuel gas pressure, gas generator speed and turbine temperature. A running trip would be accompanied by our computer data logging record of all other temperatures, pressures, flows and speeds. By this means we were able to build up a picture of the events leading to a failure and embark on a course of action.

Each trip attributed to the machinery was then recorded to show how the reliability was proceeding in relation to the targets. The cusum chart was used to provide a visual indication of our progress. Figures CS2.4 and CS2.5 show the start and run reliability plots for one machine. The reliability of the unit between any two points is indicated by the slope of the line joining them. The chart is arranged so that a horizontal line is on target, an upward slope is better and a downward slope is worse than target.

6.2 In service result

The compressor vibration levels were maintained at an acceptably low level, but the main seal oil supply pump, being mechanically driven from the power turbine shaft,

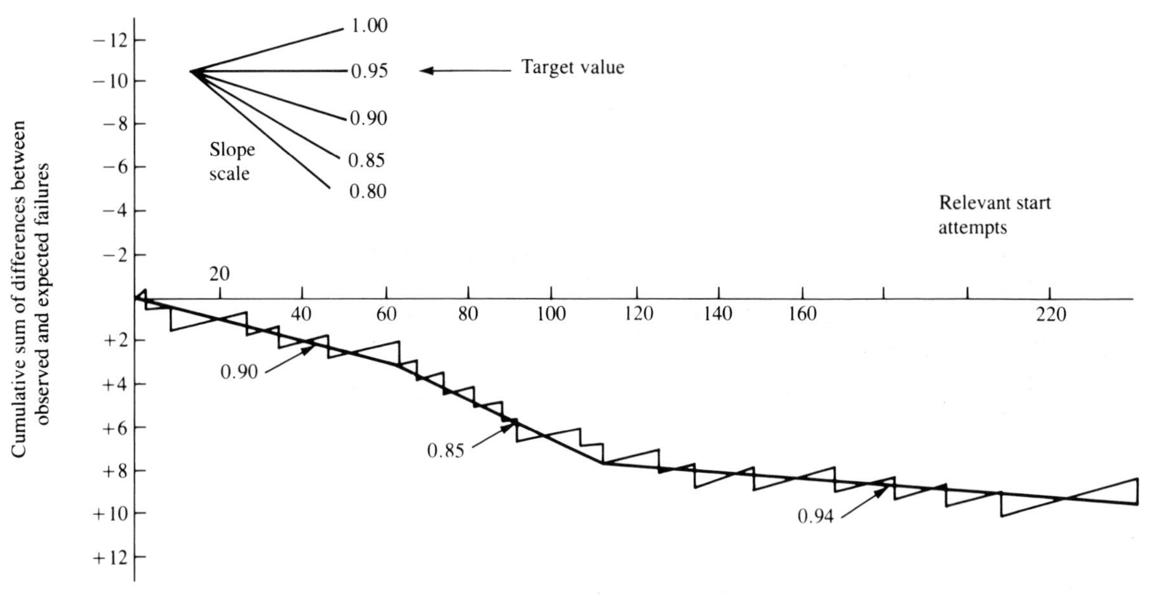

Fig. CS2.4. Start reliability cusum plot

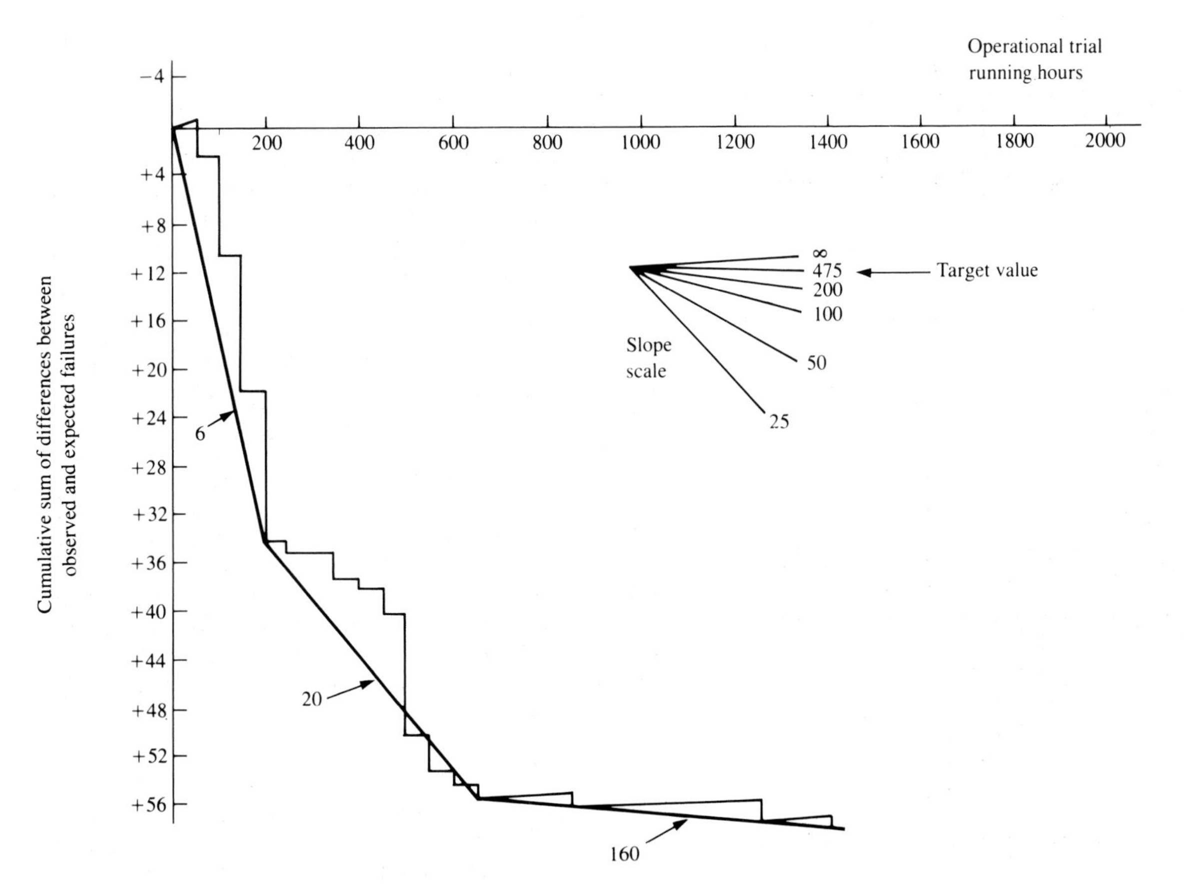

Fig. CS2.5. Run reliability cusum plot

was unable to satisfy the greater flow requirement at low speeds. We were forced to introduce an operational restriction to avoid the low speed range, until a suitable pump could be identified and installed.

Our cusum charts for the trial (Figs CS2.4 and CS2.5) show that, initially, the start reliability plots were close to target. However, after about 50 starts the start reliability of both machines began to deteriorate. The run reliability started well below target to the extent that it was very soon realised we would have great difficulty in achieving the target figures within the operational trial period.

Analysis of the particular trips causing these reductions indicated two problem areas. The fuel gas control system was leading to inconsistent ignition and light-off during the gas generator startup. Under running conditions the compressor lubrication and seal oil system was suffering transient pressure dips, especially when accelerating to full speed.

Comparison of our start records from the chart recorders revealed that fuel gas ignition did not always occur at the same gas generator speed. The period between 'ignitors-on' and 'light-off' also varied, and no obvious trend difference existed between the new and refurbished gas generators. The fuel control system rather than the gas generators was believed to be the source of the problem. Analysis of the chart recorder traces highlighted a number of changes necessary in the control parameters and the way in which control was being exercised.

The compressor lubrication and seal oil system provided the major contribution to the low running reliability. Initially we believed that our original bushing ring problem was causing pressure fluctuations due to difficulties in controlling the higher oil flow rates. To some extent this was true, but we also highlighted another underlying problem associated with the supply pump control. It was stated earlier that the main supply pump was mechanically driven from the power turbine. Because the output of this type of pump is proportional to engine speed, it is incapable of delivering sufficient flow at low speeds. To overcome this deficiency, the design had included an electrically driven auxiliary pump to cover the startup phase, which automatically switched off after a certain speed and allowed the main pump to take over supply. Despite a considerable overlap between the two pumps, the system was experiencing a large pressure drop as the electric pump shutdown. Detailed investigations of flows, pressures and the control of spill back valves eventually provided an answer, and changes to pipework and valves were required.

6.3 *Presentation of results*

Throughout the operational trial, the machinery supplier participated in the exercise of recording and analysing data. The factual aspect of the resulting reliability figures and the obvious trends produced by the cusum charts, served as a permanent indicator of performance, and amplified the need to improve matters. With the aid of the reliability programme we were able to demonstrate the persistence of certain problem areas, and the inconvenience and cost of an unreliable machine. It highlighted to the compressor vendor, the dominance of the large number of seal oil system control trips in comparison to the rest of the machinery train.

It was perhaps this latter aspect that also encouraged him to produce an improvement to the shaft dynamics. By introducing an improved bearing design, we were able to return to the original two bushing ring arrangement, with the lower oil flow requirement. This in turn voided the need to install larger pumps and produced a much improved and consistent seal oil control.

The cusum chart for run reliability, (Fig. CS2.5) indicates the cumulative effects of these modifications with a significant improvement after 600 hours.

7.0 Review of results

7.1 *The design assessment*

Predictions for the start reliability provided a good forecast, with both units meeting their target figures by the end of the operational trial. Greater concentration on achieving the correct starting logic and fuel gas setting-up procedures probably helped to achieve this.

The final run reliability figures were significantly short of the predictions. The changes recommended by the design assessment were of little consequence in terms of the eventual overall reliability, although component improvements based on experience from other compressor stations were successful.

The difficulty in assessing the small improvements was masked by the more major design problems. However, it is more constructive to state that the design assessment did not produce more unreliability by its recommended changes and verified the existing designs wherever any meaningful investigation was conducted. It is an important fact that throughout the operational trial we suffered few unexpected component failures, despite the many fundamental design problems which imposed a more arduous operational condition on many of the systems.

We established a number of areas where stricter quality control had to be exercised, but any true improvements gained are difficult to quantify.

7.2 *The operational trial*

Despite the initial good start reliability the figure soon reduced to 0.85 over the next 100 starts. We experienced inconsistent starting and instituted improvements in the logic control and hardware associated with the fuel gas system. At the end of the trial we concluded that both machines had effectively reached the target of 0.95.

The run reliability, as stated earlier failed to achieve the predicted targets due to the mechanical design and control weaknesses of the compressor. The degree of unreliability however must be shown in the correct context.

Detailed analysis showed that most trips occurred

either after initial start up or just before shutdown during a pump changeover phase. Because our operating pattern was more transient than expected, especially with the large number of short test runs required, we recorded a high number of failures over relatively few running hours. This yielded an exceptionally low MTBF. A requirement for long duration runs would have produced very different results.

Another distorting factor was that the compressor created a more arduous set of operating conditions than expected for many of the associated components and systems. Not only were we limited in operational terms to avoid certain known problem areas, we also conducted many unsuccessful tests and introduced minor modifications which in turn imposed a strain on other equipment. Failed test runs produced more shutdown trips, and changes to the seal oil system often adversely affected the lubrication system and vice-versa. It, therefore, may be argued that we often logged more failures than were strictly due to the original equipment design.

Throughout the programme we were continuously attracted to the possibilities of subtracting the compressor figures from the final data. We had adopted this process for revealing secondary problem areas and developing a priority listing for other investigations. It soon became apparent we could also use this technique for presenting a much improved, and perhaps more accurate reliability picture, for the rest of the machinery. Without the compressor problem the MTBFs of the units are in excess of the target value, 600 hours for one unit and greater than 800 hours for the other. Unfortunately it had the effect of distracting our efforts from the principal compressor problem, having derived a false sense of achievement. We therefore continued to report overall reliability and show the machinery package in its worst light.

7.3 *The achieved reliability*

At the end of the operational trial the reliability values read from the cusum charts were as shown in the table below.

	Number of starts	*Start reliability*	*Hours run*	*Run reliability (MTBF)*
A unit	270	0.94	1600	160 hours
B unit	274	0.98	2271	120 hours

The start reliability of *A* unit is considered acceptably close to the target.

The run reliabilities at the end of the trial were totally determined by the compressor problems.

8 Conclusions

The reliability prediction did not match the in-service running performance over the operational trial in terms of MTBF. Its inability to do so was not directly the result of a poor reliability assessment on the area considered, but due to the assumption that the basic design was correct and that design errors if any could be identified and corrected during the early stages of the trial. Our assumption may have been realistic had it not been for the compressor problems. With the advantage of hindsight we might have amended our assessment and target figures to reflect the use of new equipment. There is no doubt that the compressors severely reduced the reliability of the machinery train, but it would be incorrect to conclude that the reliability assessment and its forecast were meaningless.

Initially the compressor accounted for 60 to 70 per cent of the unit unreliability. However, there was still appreciable debugging of minor design problems on the rest of the unit at this stage. By the end of the trial the compressor was accounting for virtually 100 per cent of the unreliability.

A realistic operating trial must continue for an adequate number of hours in order to derive a satisfactory assessment. Experience gained during the operational trial indicates that debugging minor design problems can take up to 500 hours. We had selected 3000 hours for each unit over a period of one year, but actually only achieved 1500 hours on one machine and 2000 hours on the other in 27 months. Much of our running was restricted to short periods which produced a very low estimate of run reliability.

Despite the transient nature of running, the principal benefits derived from the reliability programme arose during the operational trial, where clear identification of problems often led to immediate corrective action. Whilst design effort was concentrated on the compressor, the more detailed analysis of operational data identified underlying trends and secondary problem areas, which were often more amenable to swift remedial action. The implementation of the start recorders and data logging system can be immensely beneficial tools in the fault diagnosis process.

The final and overriding conclusion is that fundamental design problems on recently developed equipment can dominate the reliability of the complete machinery system. To highlight and overcome this aspect in the reliability assessment stage, even if possible, would require a far more detailed review of critical and unproven areas than a statistically based reliability prediction.

CASE STUDY No. 3. AVAILABILITY ASSESSMENT FOR AN AMMONIA PLANT

T. R. Moss   RM Consultants Ltd

1. Introduction

AECI Ltd requested an availability assessment for a new ammonia plant currently under design. The plant was a large (1000 tonnes/day) single-system unit where high levels of availability were essential for economic operation. The objectives of the study were to identify the probable loss of production due to plant unavailability and to identify critical areas of plant and equipment.

2 Assessment procedure

The ammonia plant assessment followed the general form of procedure discussed in Part Four, section 11.3. Although this was shown as a series of discrete steps there was, of course, considerable interaction particularly in the first four stages. The main parts of the study are discussed below.

3 Definitions

The assessment objectives agreed between AECI Ltd and NCSR included the following.

(1) Estimates of the product-loss to be expected from the unavailability of sub-systems, main plant items and the plant as a whole.

(2) The identification of areas and operating methods which are critical to the availability of the plant.

(3) The identification of areas where further, more detailed study may be required.

Plant boundaries were also defined and included a clear statement of the assumptions to be made in the assessment on the inputs and outputs across these boundaries. In addition to the primary feedstock of coal, inputs of various sources of water, electric supply, and important chemicals were considered. The outputs from the plant included desirable byproducts such as liquid oxygen and carbon dioxide, as well as ammonia. Undesirable outputs also needed consideration and included contaminated nitrogen and methanol. Ash from the coal burned in the gasifiers constituted a major undesirable output.

It can be appreciated that unavailability of input feedstock or the inability to process outputs can also cause the plant to be shutdown. These factors can easily be forgotten and need to be noted at an early stage in any assessment.

The failure states and availability model required special consideration in this study because of the requirements of AECI Ltd. Because there was component redundancy in certain areas it was possible to postulate a wide range of operational states. However, during the preliminary appraisal it became evident that the ability of the system to operate below 100 per cent output was largely determined by the availability of one of the major compressor stages. The following operational states were, therefore, chosen.

(a) 100 per cent output.

(b) 50 per cent output.

(c) 0 per cent output.

Within these definitions failure criteria were established to meet the operator's specific interest in product unavailability predictions as follows.

(1) Reduced-output unavailability – failure to maintain 100 per cent output (that is, probability of being in states other than state (a)).

(2) Shutdown unavailability – failure to maintain 50 per cent output (that is, probability of being in state (c)).

It should be noted that although *the defined operating states are mutually exclusive the failure states are not.*

The availability model was based on the above failure criteria. Consider a minor rearrangement of the general expression for availability

$$A = 1 - D$$

The plant availability can be seen as a function of the three operating states, viz

$$A_p = \{P(a) \times 100\%\} + \{P(b) \times 50\%\} + \{P(c) \times 0\%\}$$

where $P(a)$, $P(b)$, and $P(c)$ are the probabilities of the system being in one of the states a, b, or c, respectively, and

$$P(a) + P(b) + P(c) = 1$$

In terms of the specified failure criteria the system product unavailability can also be represented by

$$D_p = \frac{D_{RO} + D_{SD}}{2} \quad \text{since} \quad D_{RO} = 1 - P(a) \quad \text{and} \quad D_{SD} = 1 - P(a) - P(B)$$

D_{RO} and D_{SD} are the unavailabilities calculated from the general expression

$$D = \frac{\lambda T}{1 + \lambda T}$$

where mean availability, A, is defined as

$$A = \frac{\text{MTBF}}{\text{MTBF} + \text{MTTR}}$$

where

MTBF = Mean time between failures
MTTR = Mean time to repair

and the complement of mean availability, the mean fractional dead time, D, represents the proportion of time spent by the plant in the failed state, and is defined as

$$D = \frac{\text{MTTR}}{\text{MTBF} + \text{MTTR}}$$

$$= \frac{\lambda T}{1 + \lambda T}$$

where

λ = Mean failure rate (= 1/MTBF)

T = Mean repair time

It follows that

$$A + D = 1$$

In this context, therefore, availability and fractional dead time have the characteristics of probabilities.

4 Plant description

The plant is essentially a single-stream system involving 22 units each of which have some component redundancy. A simplified flowsheet is shown in Fig. CS3.1. Raw coal is fed to the steam-raising plant and the coal preparation plant. In the latter the coal is pulverised and the coal dust then passed pneumatically into the gasifiers where it is mixed with oxygen (from the air separation plant) and steam. The mixture is burned producing raw synthesis gas comprising CO, H_2, and CO_2 with small quantities of impurities. The dust is removed from the gas in disintegrators and electrostatic precipitators before the first stage of compression.

After compression the raw syngas passes through a complex clean-up process to remove impurities and CO_2, and then pure nitrogen is added to attain the correct stoichiometric composition. The second stage of compression involves a multi-stage centrifugal compressor driven by two steam turbines in tandem. In the first three stages the gas is compressed from 45 to 212 bar. Before entering the fourth stage the gas is joined by recycled gas from the synthesis unit and the combined gas streams are then compressed to 226 bar and passed to the synthesis unit.

In the synthesis unit the gas mixture, at high temperature and pressure, is converted into ammonia in catalyst beds. The ammonia is separated from the unsynthesised gas by cooling against cold ammonia and the unsynthesised gas recycled to the last stage of the syngas compressor. The ammonia is reduced in temperature in a flash drum before passing to the refrigeration unit.

The refrigeration unit cools the product to −37°C so that it can be stored at low pressure. It also provides

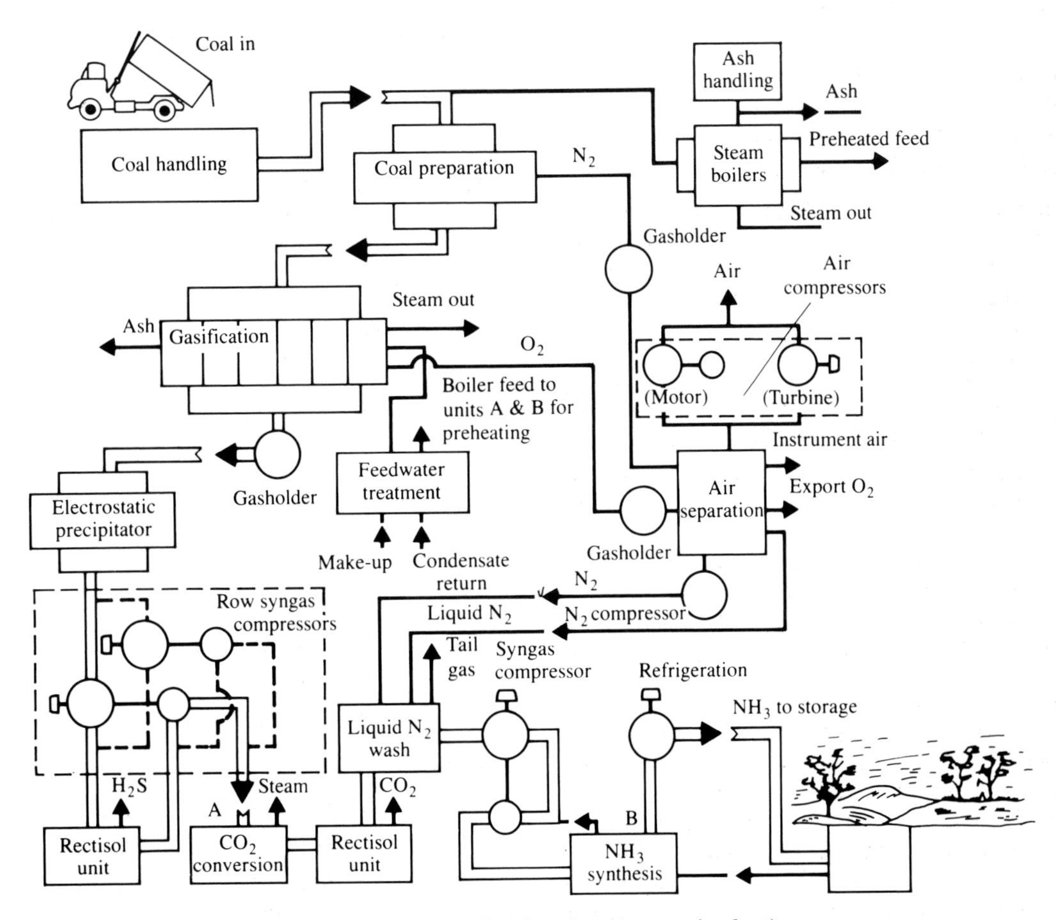

Fig. CS3.1. Simplified flowsheet (ammonia plant)

cooling for parts of the clean-up system. It differs from conventional refrigeration plants in that the refrigerant does not operate in a closed loop; the refrigerant being, in fact, the process gas.

There are a number of ancillaries associated with the plant; for example, conventional steam boilers, cooling water and air-separation units, and so on. Although these are not directly involved in the main stream process, their operation is an essential part of the total system. As such their availabilities must also be considered in the assessment of plant availability.

5 The assessment process

Assessment involved developing block diagram and failure logic for the reduced-output and shutdown cases from the simplified flowsheet and *P* and *I* diagrams. The block diagram for the plant and failure logic for the shutdown case are shown in Figs CS3.2 and CS3.3.

Individual units were considered in the same way as the plant although, of course, it was not always necessary to develop the failure logic in diagrammatic form. In this assessment it was possible to calculate the availability of most of the units from tables developed directly from the process and instrumentation flowsheets. An example from the assessment of the raw synthesis gas compressor unit is shown in Tables CS3.1 and CS3.2.

6 Estimating failure rates

Of the reliability parameters the failure rate is most prone to variation. It is important that the consequences of failure are fully explored in conjunction with information on the failure modes associated with particular faults. On mechanical valves, for example, the major failure mode considered was external leakage; the consequence of such a leak was obviously different for flammable synthesis gas and water. Estimates of the appropriate failure rate may, therefore, vary quite considerably from unit to unit.

At the same time it was thought worth sacrificing some degree of accuracy in the interest of simplicity. In this case it was estimated that the variation in operating conditions was unlikely to give data variations much

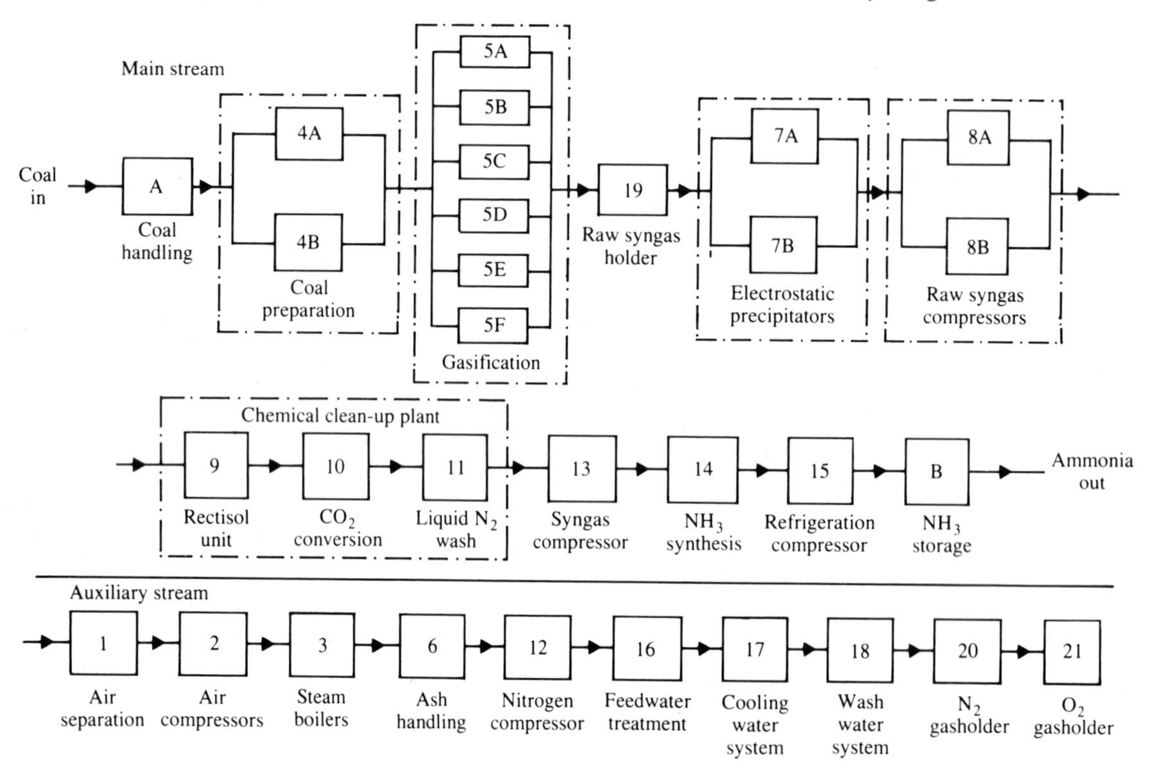

Fig. CS3.2. Ammonia plant (block diagram)

Table CS3.1. Raw-synthesis-gas compressor unit

Equipment	*No. off*	*Failure rate (f/y)*	*Average repair time (h)*	*Redundancy*		*Average failure rate*		*Average dead-time*	
				Reduced output	*Shut-down*	*RO (f/y)*	*SD (f/y)*	*RO (h/y)*	*SD (h/y)*
Raw-syngas compressor set	2	8.6	17.1	2/2	1/2	17.28	0.14	295.5	2.41

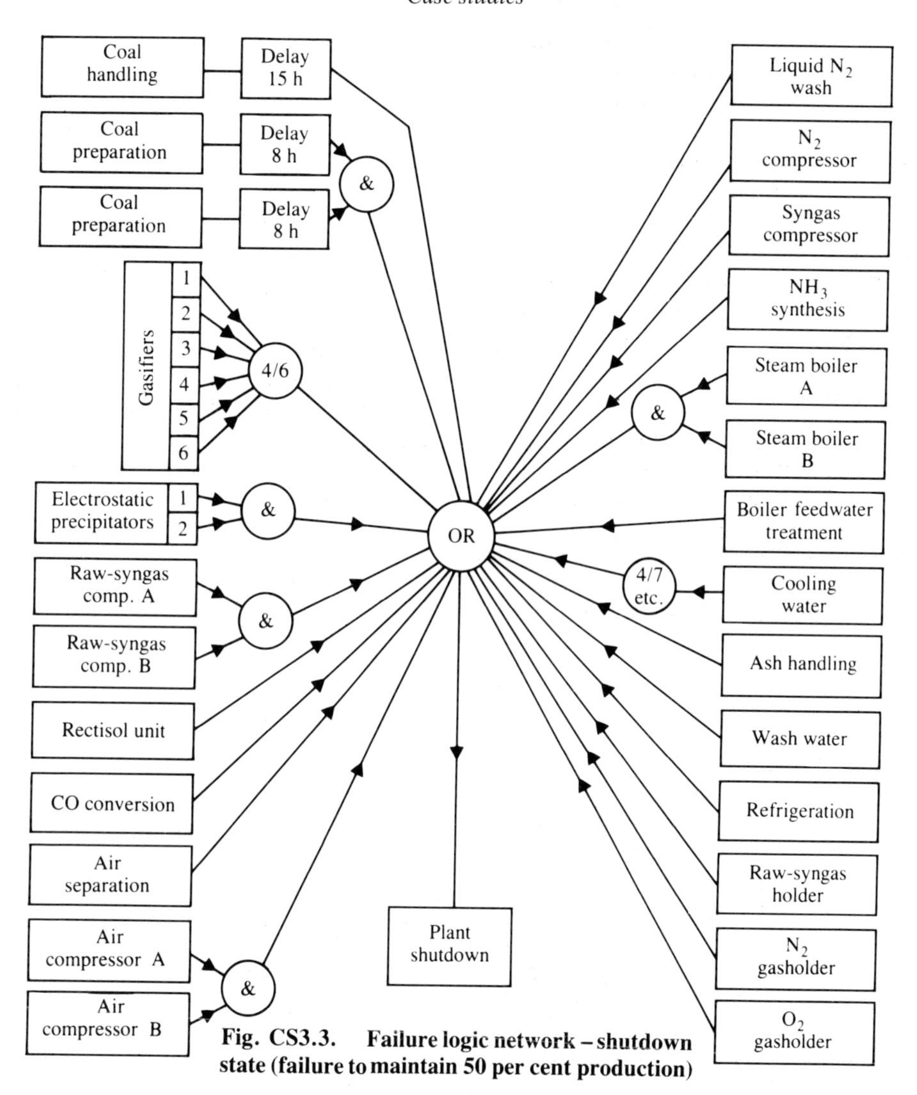

Fig. CS3.3. Failure logic network – shutdown state (failure to maintain 50 per cent production)

Table CS3.2. Single compressor set. Drg. XYZ

Equipment	No. off	Failure rate (f/y)	Average repair time (h)	Redundancy: Reduced output	Redundancy: Shut-down	Average failure rate: RO (f/y)	Average failure rate: SD (f/y)	Average dead-time: RO (h/y)	Average dead-time: SD (h/y)
Steam turbine	1	0.6	70	1/1	1/1	0.6	0.6	42	42
Compressor casing	3	0.54	30.5	3/3	3/3	1.6	1.6	49.4	49.4
Lubrication system	1	1.07	4.8	1/1	1/1	1.1	1.1	5.1	5.1
Gas coolers	5	0.01	72	5/5	5/5	0.05	0.05	3.6	3.6
Ejector	2	0.1	24	1/2	1/2	negl	negl	negl	negl
Pneumatic valves	4	0.35	4	4/4	4/4	1.4	1.4	5.6	5.6
Pressure-relief valve	3	0.5	4	3/3	3/3	1.5	1.5	6.0	6.0
Hand valves	40	0.02	4	40/40	40/40	0.8	0.8	3.2	3.2
Non-return valves	2	0.5	4	2/2	2/2	1.0	1.0	4.0	4.0
Gear seal	1	0.6	48	1/1	1/1	0.6	0.6	28.8	28.8
			17.1	Total		8.6	8.6	147.8	147.8

The figures in the redundancy columns show the ratio of equipments-required to equipments-available to maintain operation above the levels required for reduced output or for shutdown condition.

greater than a factor of two or three, and the benefits of using one data set for application to all units outweighed the other factors, at least for the first estimates.

An extract from the table of reliability parameters used in the assessment is shown in Table CS3.3.

7 Plant availability calculations

Aspects of redundancy and storage capacity were taken into account in the individual subsystem assessments. The final calculations then involved a simple summation of the downtimes for each of the defined failure states, however, expressions can be developed directly from the failure logic diagrams if preferred. To illustrate the point, consider the calculations for Unit 8 – the raw syngas compressors on which data are presented in Tables CS3.1 and CS3.2. Sections of the plant logic networks concerned with the raw syngas compressor for the reduced-output and shutdown states are shown in Figs CS3.4(a) and (b).

Either compressor failing produces a reduced-output state (Fig. CS3.4(a)) and the fractional dead time for Unit 8 is then

$$D_{RO8} = D_A + D_B$$

where

$$D_A = D_B = \frac{\lambda T}{1 + \lambda T}$$

Here

$$\lambda = 8.6 \text{ faults/year}$$

$$T = 17.1 \text{ hours}$$

$$\lambda T = 8.6 \times (17.1/8760) = 0.017$$

Therefore

$$D_{RO8} = 2(0.017)$$

Table CS3.3. Reliability data used in assessment

Equipment	*Average failure rate (f/y)*	*Average repair time (h)*
Heaters, electric	0.02	72
Ion columns	0.1	24
Motors, electric		
Small	0.03	4
Large	0.12	148
Pressure vessels	0.001	72
Pumps		
Centrifugal and general	2.6	24
Oil	0.5	8
Scrapers	0.7	8
Stirrers	0.03	4
Tanks	0.001	72
Transformers, high voltage	0.003	24
Turbines		
Nitrogen	0.4	70
Steam	0.6	70
Valves		
Hand-operated	0.02	4
Motorised	0.05	4
Non-return	0.5	4
Pneumatic	0.35	4
Pressure relief	0.5	4
Solenoid	0.08	4

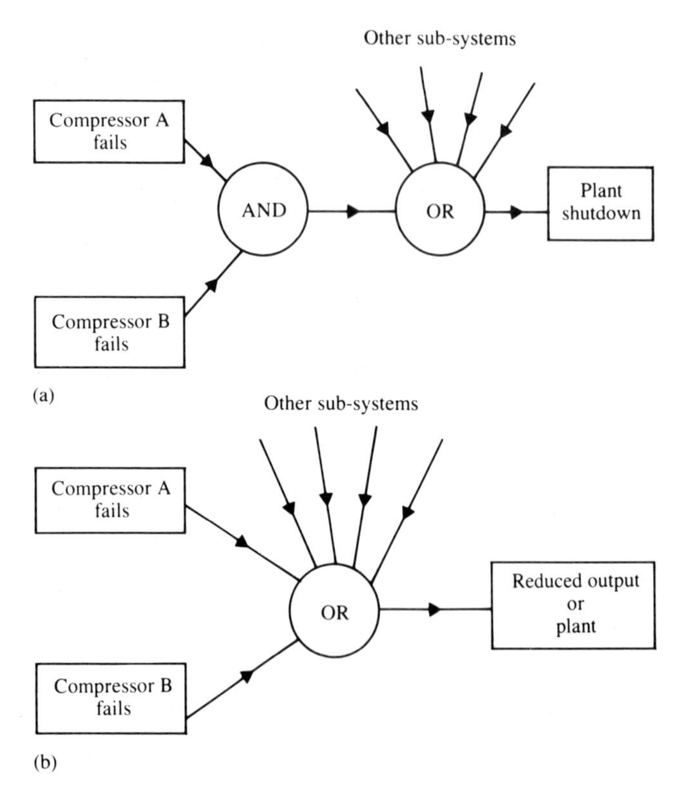

Fig. CS3.4. Logic network: (a) reduced output, (b) shutdown

The reduced-output deadtime is, therefore

$$D_{RO} = 2(0.017) \times 8760$$
$$= 295 \text{ hours per year}$$

(cf Table CS3.1).

Both compressors need to fail to produce a shutdown state (Fig. CS3.4(b)) and the fractional dead time for Unit 8 is then

$$D_{SD8} = D_A \times D_B$$
$$= (0.017)^2$$

The shutdown deadtime in this case is

$$D_{SD} = (0.017)^2 \times 8760$$
$$= 2.4 \text{ hours per year}$$

Having synthesised the deadtimes of each sub-system from component data in this way it was then possible to generate an overall availability characteristic for the plant. This involved the summation of sub-system deadtimes to give plant reduced-output and shutdown fractional dead times. Applying these figures in the simple expression developed in the section 3 gave the equivalent product unavailability.

Having synthesised the fractional dead times for each unit in the two failure states the plant availability equations involve a simple summation of the fractional dead times for the 22 units. The equation for the plant shutdown-state is defined by the logic diagram shown in Fig. CS3.3, the OR box indicating a summation of all unit fractional dead times in that failure state, viz

$$D_{SD} = D_{SD1} + D_{SD2} + \cdots + D_{SD22}$$

(the above expression is a simplification and ignores the effect of simultaneous sub-system failures).

A similar calculation is made for the reduced-output state fractional dead time (D_{RO}). The product unavailability is then defined by the equation in section 3, namely:

$$D_p = \frac{D_{RO} + D_{SD}}{2}$$

and product availability

$$A_p = 1 - D_p$$

The predicted product availability was in excess of 85 per cent. An indication of the sub-system and components which could be expected to contribute most to the overall deadtime can be seen in Table CS3.4.

Sensitivity analysis was then carried out on each of the sub-systems in turn to evaluate the effect of various stressing factors on component failure and repair characteristics. The factors considered in the sensitivity analysis included the following.

(1) Operational conditions (temperature, pressure, flow, etc.).

(2) Environmental conditions (vibration, dust, humidity, etc.).

(3) Material properties (flammability, corrosion, compatibility, etc.).

(4) New technology.

(5) Human factors.

Table CS3.4. Distribution of component and sub-system dead-times (hours/year)

	Rotating machinery			*Valves*								
Unit	*Turbines*	*Compressors*	*Other*	*Pneumatic*	*Hand*	*PRV*	*NRV*	*Heat exchangers*	*Boiler units*	*Estimated hours lost*	*Estimated days lost*	*% of total estimated lost days*
03 Steam boilers									175.2	175.2	7.3	15.5
13 Syngas compressor	84	43		4.2	1.8	1.2		2.9		137.1	5.7	12.1
08 Raw-syngas compressors	42	60		5.1	3.3	0.6	0.4	3.6		115.0	4.8	10.2
12 N_2 compressor	42	38	29		0.6	4.0	2.0	2.9		118.5	4.9	10.4
15 Refrigeration	42	38		12.0	2.1	12.0		12.5		118.6	4.9	10.4
09 Rectisol			3.5	19.9	11.5	4.8	8.0	40.0		87.7	3.7	7.9
14 NH_3 synthesis				12.6	1.8	0.2		58.5		72.9	3.0	6.5
02 Air compressors	21	22.5	23	2.8	1.7	1.0		1.5		73.5	3.1	6.6
01 Air separation			2.7	32.2	7.7	3.6	3.6	2.1		51.9	2.2	4.7
10 00 Conversion				0.9	2.8	2.0	4.0	36.0		45.7	1.9	4.1
Estimated hours lost	231	201.5	58.2	89.7	33.3	29.4	18.0	159.8	175.2	896.1		
Estimated days lost	9.6	8.4	2.4	3.7	1.4	1.2	0.8	6.7	7.3		41.5	
% of total estimated lost days	20.4	17.9	5.1	7.9	3.0	2.6	1.7	14.3	15.5			88.4
	43.4			15.2				14.3	15.5			
	88.4											

The results of this part of the exercise were incorporated into individual sub-system reports where specific recommendations as to the sensitive areas, operating procedures, levels of redundancy, spares holdings, and so on, were made, in addition to predictions of failure frequency, deadtime, and availability. The most important points emerging from these studies were also emphasised in the main body of the assessment report.

Before discussing the conclusions and recommendations with AECI engineers attempts were made to compare availability estimates with operating experience in other ammonia plants. One report in particular of experience from the operation of 23 plants in the USA generally confirmed a number of the predictions of sub-system deadtimes and lent weight to the conclusions put forward in the draft assessment report.

The draft report was discussed on site with AECI engineers and their local and plant experience incorporated to generate best estimates of plant and sub-system availabilities. Action was then taken by the company in the areas they considered would give the most cost effective results. Further assessment work was also commissioned on parts of the instrumentation.

8 Conclusions

Early results of experience with the operation of the plant are encouraging. It was notable that several of the areas identified as likely to be sensitive have, in fact, been shown to be so. In particularly, it was predicted that the steam boilers would give the highest contribution to unavailability and results show close agreement between predicted percentage of total hours lost and those obtained in practice. Similar agreement has been evident in several other areas.

The technique does, therefore, appear to be viable for the assessment of chemical plant even though many simplifying assumptions have to be made. Plant availability assessment is no substitute for good engineering design, but it does clearly add another important dimension to the information available for decision making. By concentrating on the sensitive areas one can optimise the cost effectiveness of design, operation, and maintenance effort.

CASE STUDY No. 4. RELIABILITY DATA COLLECTION AND ANALYSIS PROJECT

T. R. Moss *RM Consultants Ltd*

1. Introduction

Comprehensive reliability data are required in offshore engineering projects to support safety and availability studies. To meet this need the OREDA (Offshore Reliability Data) project was conceived in 1980 to collect, analyse, and present reliability information in the form of a Reliability Data Handbook for the benefit of the offshore industry. Funding for the data collection and analysis was provided by eight companies:

Elf Aquitaine Norge A/S
A/S Norske Shell
Norsk Hydro A/S
Statoil
Norsk Agip A/S
Total Oil Marine plc
BP Development, Norway
Saga Petroleum A/S

The main contractor was Veritec, the independent consultancy group set up by Det Norske Veritas 1984.

Failure data and engineering information used in the preparation of the OREDA handbook were collected from maintenance and operating records of the participating companies. Extraction of the reliability data was clearly never envisaged when the record systems were introduced initially; hence, data collection involved distillation of the required information from a wide variety of dispersed manual and computer files. These data were subsequently screened and processed into a format suitable for statistical analysis by the main contractor.

This case study discusses the organization and execution of OREDA data collection by one of the sub-contractors (RM Consultants Ltd) and some of the problems that needed resolution before the processed data could be transmitted for incorporation into the OREDA Handbook.

2 Organization

The specification of items on which data were to be collected was broadly agreed by the main contractor and the participating company representative. Subsequent data collection and initial data processing was carried out by consultants chosen by the individual companies.

The consultants chosen were experienced reliability engineers with considerable experience in data collection from industrial sites. In our case the project was supervised by a principal engineer with many years experience in data bank operation. Data collection was carried out by mature engineers with a background in process plant operation.

For this project, where the main office was a long way from the site, it was obviously essential to choose experienced self-motivated engineers and to support them with special training. We instituted a series of short sessions to familiarise people with the guideline documentation issued by the main contractor and to define methods and responsibilities for collecting the data before starting the main exercise.

3 Initial programme

This initial programme involved the following.

(1) Discussion of the objectives of the study and likely problem areas, including allocation of responsibility for certain aspects, for example, instrumentation.

(2) Review of Work Reports 1 and 2 (Taxonomy and Boundary specifications) including working through specific examples from typical flowsheets, *P* and *I* diagrams, and previous offshore reliability assessments.

(3) Evaluation of Work Report 3 (Failure mode specifications) involving comparisons with previous offshore reliability assessments.

Each discussion session took about four hours and resulted in a range of queries and recommendations which were transmitted back to the main contractor. The second stage was concerned with evaluation of the data collection guidelines issued by the main contractor as Work Report 4. The stages in this evaluation involved:

(4) A review of the inventory and event data collection forms including working through a number of examples.

(5) Consideration of masks to facilitate transfer of the raw data into a computerised data base for subsequent processing. In fact this proved impossible and it was necessary to devise an intermediate form for data input (Fig. CS4.1).

4 Preliminary site visit

The third stage involved a preliminary visit to the site of the participating company to establish contact with the site personnel and determine the availability of data. The objectives of the OREDA exercise and an outline of information required were discussed with the Chief Maintenance Engineer who delegated responsibility to engineers in his group for specific areas of data. These engineers, predominantly from the maintenance planning department, were introduced to the data collection team and the objectives and benefits to the company discussed. It was pointed out by the Chief Maintenance Engineer that maintenance of offshore production facilities must always remain the first priority, however, within this constraint full cooperation would be given,

XX023	HARIS–OREDA FAILURE EVENT RECORD	*002064*
AX001	*3.1.2.2.3* XO24 *04*	Taxonomy and Platform Code
BX002	*MAIN POWER GENERATION > 7 MW*	Item Description
CX003	*ELEC GEN MAIN*	System of which item is part
DX004	*A*	Failure Mode
EX005	*1*	Severity Code
FX006	*R*	Operational Mode (C, S, R, O,)
GX007	*210882*	Date of Failure or Repair D.M.Y.
HX008	*5 HOURS*	Active Repair Time
IX009	*COIL CONTACTOR*	Component or Sub-system
JX010	*BURNT OUT*	Component Failure Mode
KX011	*C*	Component Severity Code (C, D, I, U,)
LX012	*ROLLS ROYCE AVON 1533*	Manufacturer and Type
MX013	*REPLACE CONTACTOR COIL*	Event Description
NX014	*10.2 (UNKNOWN)*	Failure Cause
OX015	*STARTER MOTOR FAIL TO ENERGIZE*	Failure Effect
PX016	*COIL OF CONTACTOR*	Defective Part
QX017	*REPLACED CONTACTOR*	Repair Action
RX019	*EQUIP HIST W41/83 CH 199677*	Data Source with Card and Page
SX020	*G1010*	Tag Number
TX021	*E0101*	EDCC Code
UX022	*JAB 291183*	Analyst Initials and Date
VX018		Comments. Use 2nd line if necessary
		Final Character ?? (end of record)

Fig. CS4.1. Event input form

including access to all records available. By and large this commitment was fully honoured by the maintenance departments.

In conjunction with the key maintenance personnel we identified the sources of inventory and failure event data and took examples of print-outs and ther information away for more detailed consideration in our offices. To investigate the scope for extracting subsets of data from the computerised record system a short series of requests were discussed and specified to the data processing engineer. It transpired that the computer programs were quite inflexible and the only viable approach within the timescales available was to extract all of the data manually.

5 Main data collection

About two weeks transpired between the preliminary visit and the main data collection phase. Psychologically, spending only one or two days on site initially is likely to be beneficial, because it gives people time to adjust to a new situation. By the second visit contact had already been made and data collection engineers had spoken several times on the telephone to their opposite number. We also took the precaution of writing to these key contacts outlining the programme of work for our visit and repeating our data requirements.

During the period of two weeks between the initial visit and the main data collection phase we carried out a number of sampling exercises on data collected during the initial visit. Each company's approach to data recording was quite different, of course, and posed different problems. In one, inventory data were difficult to obtain, in the second descriptions of the failure/repair events were very limited, and in the third significant differences between the manually recorded data and personal experience was noted.

A plan for data collection was devised based on

knowledge gained during the preliminary visit and the sampling exercises. It was clear from these studies that the scope for direct collection of data from individual platforms would be extremely limited. This was due to the range of items to be covered and the difficulty of getting data collection engineers on and off platforms. For this reason it was agreed that effort should be concentrated during the site visits on accumulating the maximum amount of relevant information. These records were then scrutinised in our own offices and the relevant data forms completed, before data processing, prior to transmission to the main contractor.

Taking one company as an example, it was clear that the major problem would be acquisition of inventory data and operating experience. Information necessary to complete the data form was distributed in a number of different files none of which were in any standardized format. Population statistics were particularly difficult to obtain, and in some cases (for example, instrumentation) estimates were based on review of the piping and instrumentation diagrams.

Operating hours estimates were also difficult. For some of the large rotating machines operating hours could probably have been derived from extraction of the relevant dates and times for starts and stops recorded in logbooks, but the effort required for this task would have been prohibitive. We employed data collection for other studies (monthly availability reports) where these were to hand, but in other cases estimated actual operating hours from various reports, information on major planned outages and discussions with planning staff. Failure data was available in considerable detail in the equipment histories issued monthly.

The main data collection site-visit took four weeks. We had three engineers working full time (including some evenings and weekends) and had some difficulty in getting all the accumulated paperwork – printouts, *P* and *ID*s, files, and so on, on to the plane. The effort put into planning the operation and good record-keeping paid big dividends over the following months.

6 Data processing

Since direct transfer of relevant subsets of data on to our computer was impossible we were forced to abandon computer processing. All of the data transfer from company records had to be carried out manually. We started by using the forms developed by the main contractor, but with experience were able to introduce short cuts which limited the amount of paper required and facilitated quality control. One such form is shown in Fig. CS4.2. By using this form it was possible to crosscheck on the basis of the work order, card number, equipment tag number, and event date.

Since the analyst/event number and page number of the equipment history print-out are also identified, we had complete traceability of every recorded event. Repairs are recorded in the equipment history print-out by discipline (mechanical, electrical, instrument, and general), so a single event may, in fact, be repeated on several different pages.

Quality control checks were carried out throughout the data processing phase – mainly by the principal engineer, but sometimes by one of the other data collection engineers on his colleagues' work. These checks were employed on a significant proportion of the items covered using randomly-selected samples. They comprised the following.

(1) Event data checks involving a complete traceback of sample events employed in calculating failure rates and repair times.

(2) Inventory data checks of item descriptions and populations – carried out on virtually all taxonomy class members.

(3) Operational/repair time checks to ensure that estimates could be justified and that all repair hours were recorded on the data transfer forms.

The sampling checks were specified to ensure the following.

(4) Correct transfer of numerical data.

(5) Confirmation of correct taxonomy descriptions and failure modes.

(6) No duplication of event records.

Statistical tests to confirm exponentiality were carried out on all samples using the cumulative hazard plotting method specified in Work Report No. 5. Generally this was applied without data censoring because of the limitations in terms of the number of events in most samples. Also, of course, most populations could reasonably be expected to be of mixed ages since the platforms had been in operation for many years, so the likely distribution was exponential.

Most samples, when there were a sufficient number of events, were found to be exponential. In the few cases where non-exponential behaviour was evident (generally, small populations and few events) the data were discarded. For the situation where only one or two failures were recorded an exponential characteristic was assumed.

7 Data transmission

Transfer of the data to the main contractor was made via the Data Transmission Form shown in Fig. CS4.3. Data were aggregated from the Intermediate Data Form and transferred on to this form. Calculations of failure rates and repair times were then made and entered into the appropriate columns on the Data Transmission Form. Operating hour estimates were added later as a separate exercise.

These data transmission forms were checked before sending to the main contractor. The checks covered the following.

(1) Correct transfer of numerical data;

(2) Correct allocation to the specified failure modes;

(3) Confirmation of correct taxonomy code;

(4) Calculation checks.

PLATFORM: *03* ANALYST: *J. A. BUTLER*

SOURCE: *EQUIPMENT HISTORY 1982 – WEEK 41/83*

TAXONOMY NO: *3.1.2.2.3. ELEC. GEN.-MAIN GENERATION*

WORK CARD NO.	ANALYST'S EVENT REF. NO.	PAGE OF EQ. HIST.	OREDA CODE	MAN HOURS	EVENT DATE	OP. HOURS	TAG NO.
	4049	*21*	*1A*	*11*	*2/9/82*	—	*4600*
238106	*4050*	*21*	*3G*	*4*	*27/11/82*	—	*4400*
238368	*4051*	*21*	*3G*	*10*	*4/12/82*	—	*4400*
140154	*4052*	*22*	*30*	*4*	*5/7/82*	—	*4600*
261436	*4053*	*22*	*3G*	*3*	*7/12/82*	—	*4400*
261435	*4054*	*22*	*3G*	*4*	*17/12/82*	—	*4400*
198390	*4055*	*22*	*1A*	*22*	*25/9/82*	—	*4700*
238308	*4056*	*23*	*20*	*30*	*22/11/82*	—	*4700*
192234	*4066*	*127*	*3G*	*6*	*10/7/82*	—	*4400*
237732	*4067*	*128*	*3G*	*6*	*20/10/82*	—	*ALL*
238358	*4068*	*128*	*3G*	*24*	*20/11/82*	—	—
238294	*4069*	*128*	*31*	*8*	*21/11/82*	—	—
192276	*4071*	*130*	*3G*	*4*	*28/7/82*	—	*4700*
192261	*4072*	*130*	*3G*	*8*	*15/7/82*	—	—
261369	*4073*	*131*	*3G*	*8*	*4/12/82*	—	*4400*
237738	*4074*	*131*	*3G*	*10*	*1/11/82*	—	*4500*
237833	*4075*	*131*	*3G*	*10*	*4/11/82*	—	*4500*
192286	*4076*	*131*	*3G*	*2*	*4/11/82*	—	*4500*

DATE: *9/2/84* REVISION: *1* CHECKED BY: T R Moss

Fig. CS4.2. Intermediate data processing form

DATA FORM							**OREDA**
Taxonomy No. *3.1.1.1.3*	Item Description *ELECTRICAL SYSTEM. POWER GENERATION. MAIN POWER GENERATION. TURBINE DRIVEN. INDUSTRIAL LESS THAN 7 MW*						
Make/dimension *GAS TURBINE–SINGLE SHAFT–14 STAGE AXIAL COMPRESSOR–4 STAGE TURBINE–7400 R.P.M. ALTERNATOR 3.26 MW AT 11 KV AND 50 HZ. NOT BLACK START*							Platform Id. *68*
Population (no. of units) *3*			Time in service (10E6 hours) Calendar time *0.0987*		Operational time *0.0929*		No. of cycles/demands *1365*
Failure mode code	No. of failures	Active repair time (hours)	Repair time (manhours) Aggregated	Range max/min	Failures per 10E6 hours Calendar time	Oper. time	Failures per 10E6 demands
1.A	*1*		*1*		*10.1*	*10.8*	*733*
1.B	*13*		*33*		*132.0*	*140.0*	
1.C(a)	*4*		*13*		*40.5*	*43.1*	
1.C(b)	*1*		*24*		*10.1*	*10.8*	
1.0	*13*		*38*		*132.0*	*140.0*	
2.D	*1*		*120*		*10.1*	*10.8*	*733*
2.E	*1*		*36*		*10.1*	*10.8*	
2.0	*13*		*71*		*132.0*	*140.0*	
3.0	*137*		*644*		*1390.0*	*1475.0*	
4.J	*10*		*92*		*101.0*	*108.0*	
All modes *194*		*5.5*	*1072*	*1–120*	*1970*	*2088*	*1466*

Comments	%		%
FUEL SYSTEMS	*14*	*LUBE SYSTEM*	*8*
BURNERS	*9*	*COOLING SYSTEM*	*7*
CONTROL	*3*	*FLAME DETECTION*	*10*
INSTRUMENTS	*5*	*FASTENINGS*	*1*
FOULING	*8*	*STARTER DRIVE*	*1*
ELECTRICAL	*9*	*UNKNOWN*	*10*
AIR INTAKE	*15*		

OPERATIONAL TIME SCALED UP FROM RECORD OF SHORTER PERIOD

SINTEF 3–1984 Date: *5.4.84* Rev. No. *0*

Fig. CS4.3. Data transmittal form

Traceback in this case was restricted to the intermediate form.

During the data processing phase queries which could not be resolved on the telephone were recorded. These queries were resolved in subsequent visits to site.

8 Reporting

Reporting by the sub-contractor needed to be to the main contractor and the participating company representative. The main contractor was provided with monthly progress reports and a full set of completed Data Transmission Forms for individual platform samples in each taxonomy class. Additionally the participating company representative received a final report for each taxonomy class containing commercially-sensitive information on equipment manufacturer, type, and so on, and more detailed information on major failures, maintenance and testing, operational conditions, and experience. An example of a typical Progress Report is shown in Fig. CS4.4.

9 Conclusions

Overall the data collection proved a difficult but rewarding task. Of course there were a number of problems which needed to be resolved. Many of these would have been trivial if the maintenance information systems had been designed to provide reliability data. That they were not is no reflection on the companies concerned, whose

RM CONSULTANTS LTD

OREDA

PARTICIPANT —

PROGRESS REPORT NO 2 — OCTOBER 1983

Time spent this period — 540 hours
Total time spent — 675 hours

Expenses incurred this period — 2699.70
Total expenses incurred — 3399.70

Work carried out in October involved:

1. Visit by the RMC team to Aberdeen for the collection of maintenance/failure reports and inventory data.
2. Analysis of available data and compilation of equipment inventories and failure reports. This work has been limited to a few platforms in order to produce positive results and to identify possible areas of uncertainty. Examination of some areas of the taxonomy has been deferred until further inventory data can be obtained.
3. These pilot exercises have highlighted the need for further definition and interpretation of the Taxonomy, Boundary Spec. and Failure Mode Spec. The Main Contractor has been informed and detailed notes are being kept at RMC.
4. Details of progress to date are given in the Appendix attached.

T R Moss
1 November 1983

Fig. CS4.4. Progress Report example

prime interest is in maintenance, but hopefully, attention will be given to this aspect in future maintenance information system specifications. The following data proved most difficult to obtain.

(1) Inventory data – particularly item populations.

(2) Failure mode descriptions. Information recorded concerned repair operations, and in a proportion of cases it was difficult to identify the cause and effect of the failure event.

(3) Failure event populations. Repairs frequently required action from different disciplines, each covered in different parts of the record system. Repair and scheduled maintenance were frequently combined.

(4) Operating hours. Operating histories were recorded in logbooks (or logsheets) for certain machines. Even then the reasons for shutdown were not always given. The extraction of running hours from diary information is tedious and time-consuming. No logbooks were kept for other items (for example, valves), hence, operating estimates must be derived from other sources (for example, daily production telexes).

(5) Restoration times. Manhours only were recorded, and often more than one man was employed. Active repair times needed to be derived from total repair manhours. Waiting times, that is, time between shutdown and start of repair and also time between end of repair and start-up, were seldom recorded.

APPENDIX

NOTATION

A	Availability
A_s	System availability
$A(t)$	Availability function
C	Number of censored items
dt	Infinitessimally small time interval
du	Infinitessimally small interval
$f(t)$	Failure probability density function (failure p.d.f.)
$f(u)$	Probability density function
F	Cumulative failures
	Probability that an item has failed
Fc	Common cause failure probability
Fs	Probability that a system is in a failed state
$F(t)$	Probability of failure before time t
	Cumulative distribution function
$F(t_i)$	Proportion failed in sample
$\hat{F}(t_i)$	Estimate of probability of failure before time t
h_0	Sequential probability ratio test [Accept line, $T = h_0 + S\{N(t)\}$]
h_1	Sequential probability ratio test [Reject line, $T = -h_1 + S\{N(T)\}$]
H	Cumulative hazard
$H(t)$	Cumulative hazard function
$\hat{H}(t_i)$	Estimate of cumulative hazard function at observed failure time, t_i
i	Failure number
k	Total number of failures
	Proportionality constant in $Z(t) = kt$
K	MTBF multiplication factors for reliability demonstration
K_i	Stress factor for stress i
L	Load
$\bar{L}$	Mean load
LCC	Life cycle costs
m	Expected number of failures, λT
MTTF	Mean time to failure
MTTR	Mean time to repair
N	Number of items in a population
$N(t)$	Cumulative failures before time t
$N(T)$	Total number of failures
n	Number of items in a sample
	Number of units up
$n(t)$	Number of units which fail in each time period
$n(t_2 t_1)$	Number of failures occurring in the time interval t_1–t_2
$n-k$	Number of censorings (survivors)
pdf	Probability Density Function
P	Probability
	Probability that an item is working
P_F	Failure probability
P_S	Probability that a system is working
P_μ	Estimate of percentage failed at mean life
r	Number of units operating
R	Reliability
$R(t)$	Reliability function
	Probability of survival to time t
$R_s(t)$	System reliability function
s	Estimate of standard deviation
	Number of survivors
	Proportionality constant in sequential probability ratio test
S	Strength
$\bar{S}$	Mean strength
t	Time
t	Aggregate system life
t_1	Smallest ordered age at failure
t_2	Second smallest ordered age at failure
t_i	ith ordered age at failure
	Interval between $(i-1)$th failure and the ith failure
t_k	Largest ordered age at failure
T	Accumulated test time (total elapsed time)
	Cumulative time
T_j	Time at jth ordered censoring
U	Unavailability
U_s	System unavailability
x	Number of failures
y_i	System age at the ith failure
$z(t)$	Hazard Rate Function (or instantaneous failure rate)
$\hat{z}(t)$	Estimate of Hazard Rate Function
$\hat{z}(t_i)$	Estimate of hazard rate at ith failure
α	Producers risk
	Some small value
$1-\alpha$	Upper confidence limit
	High probability of accepting the product
β	Consumers risk
	Low probability of accepting the product
	Weibull shape parameter
	Common cause failure proportion of overall failures
$\hat{\beta}$	Estimate of Weibull shape parameter
γ	Weibull location parameter (alternative symbol)
η	Scale parameter for the Weibull distribution 'Characteristic life'
$\hat{\eta}$	Estimate of characteristic life
θ	Mean time between failures (MTBF)
$\hat{\theta}$	Estimated MTBF – T/x
	Current MTBF
θ_u	Upper MTBF
θ_l	Lower MTBF
λ	Failure rate
$\hat{\lambda}$	Estimate of failure rate – x/T
λ_u	Upper failure rate
λ_l	Lower failure rate
μ	Repair rate
	Mean life
σ	Standard deviation of a random variable or of a population
σ_s	Standard deviation of strength distribution
σ_l	Standard deviation of load distribution
χ^2	Chi-squared distribution

REFERENCES

(1) Carter, A. D. S. (1974) 'The bathtub curve for mechanical components – fact or fiction', *Improvements of reliability in engineering*, IMechE, London, pp. 50–58.

(2) Duane, J. T. (1964) 'Learning curve approach to reliability', *IEEE Trans. Aerospace*, **2**, 563–566.

(3) Jardine, A. K. S., Goldrick, T. S., and Stender, J. (1976) 'The use of annual maintenance cost limits for vehicle fleet replacement', *Proc. Inst. Mech. Engrs.*, **190**, 71–80.

(4) de la Mare, R. F. (1979) 'Optimal equipment replacement policies', National Centre of Systems Reliability Research Report NCSR R21.

(5) Weibull, W. (1961) *Fatigue testing and the analysis of results*, Pergamon Press, New York.

(6) Nelson, L. S. (1967) 'Weibull probability paper', *Industrial Quality Control*, **17**, p. 453.

(7) Benard, A. and Bos-Levenbach, E. C. (1973) 'Het vitzetten van waarnemingen op waarshijnlijkheidspapier' (The plotting of results on probability paper), *Statistica*, **1**, 163–173.

(8) Barnett, V. (1975) 'Probability plotting methods and order statistics', *J. Roy. Stat. Soc., Ser. C*, **24**, 95–108.

(9) Nelson, W. (1969) 'Hazard plotting for incomplete data', *J. Quality Technol.*, **1**, 27–52.

(10) Cohen, A. C. (1965) 'Maximum likelihood estimation in the Weibull distribution based on complete and on censored samples', *Technometrics*, **7**, 579–588.

(11) Carter, A. D. S. *Mechanical Reliability*, Macmillan, London.

(12) Murdoch, J. and Barnes, J. A. (1970) *Statistical tables for science, engineering, and management*, Macmillan, London.

(13) Ascher, H. and Feingold, H. (1984) *Repairable systems reliability*, Dekker, New York.

(14) OREDA (1984) *Offshore Reliability Data Handbook*, PO Box 370, N – 1322 Hovik, Norway.

(15) Reliability Analysis Centre (1986) *Non electronic parts reliability data*, Rune Air Development Centre, Griffiss AFB, New York, USA.

(16) Ministry of Defence (1983) 'MOD practices and procedures for reliability and maintainability: Part 3 – Reliability prediction', DEF STAN OU – 41 Part 3, December.

(17) Smith, D. J. (1985) *Reliability and maintainability in perspective: practical, contractual, commercial, and software aspects* (2nd Edition), Macmillan, London.

(18) O'Connor, P. D. T. (1985) Practical Reliability Engineering (2nd Edition), Wiley, New York.

(19) McCormick, N. J. (1981) *Reliability and risk analysis*, Academic Press, New York, Ch. 7.

(20) Billinton, R. and Allan, R. N. (1983) Reliability evaluation of engineering systems, Pitman, London, Chs. 8 and 9.

(21) *BS 4778: 1979 Glossary of Terms used in Quality Assurance including Reliability and Maintainability Terms*, British Standards Institution, London.

(22) *MilHdbk 472: Maintainability Analysis*, US Department of Defense, Washington, DC.

(23) Snaith, E. R. 'The correlation between predicted and observed reliabilities of components, equipments, and systems', NCSR-R18, UKAEA.

(24) Rothbart, K. A. (1964) *Mechanical design systems handbook*, McGraw Hill, New York.

(25) Green, A. E. and Bourne, A. J. (1972) *Reliability technology*, Wiley Interscience, New York.

(26) Smith, D. J. (1985) *Reliability and maintainability in perspective*, Macmillan, London.

(27) NPRD 3. *Non-electronic parts reliability data*, US Reliability Analysis Center, New York.

(28) *IEEE STD 500 – Nuclear Reliability Data Manual* (1977), IEEE, New York.

(29) Spanninga, A. and Westwell, F. (1981) 'Brent gas disposal system performance forecasts by Monte Carlo techniques', *3rd National Reliability Conference*, Birmingham 1981, Institute of Quality Assurance.

(30) Moss, T. R. and Piotrowicz, V. (1986) 'HARIS, An Information System for Hazard and Risk Studies', *EureData Conference*, Heidelberg, Springer Verlag, Berlin.